Drilling Down

To Tim —
my good friend and
colleague. I have
benefitted so much
from knowing you —

Joel

Joseph A. Tainter • Tadeusz W. Patzek

Drilling Down

The Gulf Oil Debacle and Our Energy Dilemma

Copernicus Books
An Imprint of Springer Science+Business Media

Joseph A. Tainter
Department of Environment and Society
Utah State University
Logan, UT 84322 USA

Tadeusz W. Patzek
Department of Petroleum
and Geosystems Engineering
The University of Texas at Austin
Austin, TX 78712 USA

© Springer Science+Business Media, LLC 2012

Published in the United States by Copernicus Books,
an imprint of Springer Science+Business Media.

Copernicus Books
Springer Science+Business Media
233 Spring Street
New York, NY 10013
www.springer.com

Library of Congress Control Number: 2011935227

Manufactured in the United States of America.
Printed on acid-free paper

ISBN 978-1-4419-7676-5   e-ISBN 978-1-4419-7677-2
DOI 10.1007/978-1-4419-7677-2

Acknowledgments

It is a pleasure to acknowledge the many people who encouraged us and helped with this book. Myrna and Charles Hall of the State University of New York at Syracuse suggested that we write it, and this project would not have been undertaken without their foresight. Joyce Vandewater prepared a number of illustrations, excellently as always. Steve Balogh kindly provided the data for Fig. 5.9.

Joseph Tainter is particularly pleased to acknowledge the support and sharp observations of his wife, Bonnie Bagley, who has been a true intellectual partner in all of his work.

Tad Patzek would like to thank Dr. Paul Bommer of the University of Texas, Austin, for numerous discussions, corrections, and help with the various technical aspects of the Macondo saga. He also wants to thank his son, Lucas, for ample feedback and editorial suggestions, and his wife, Joanna, for saintly patience.

Finally, it is a particular pleasure to thank Dr. David Packer for being the best editor either of us has worked with. David improved the text greatly, and if what you see here achieves its purpose, it is largely thanks to him.

Contents

Chapter 1

Introduction

We begin this book during the Fourth of July weekend, 75 days after the Deepwater Horizon exploded, burst into flames, and sank, killing 11 men. In the wake of this accident came the worst environmental disaster in U.S. history. The starting date of our writing is significant because this is a weekend when normally thousands of people would descend on the beaches and restaurants of the Gulf Coast. The Gulf is a place of great bounty. A couple of hours with some traps produces enough blue crabs to make a cauldron of gumbo that can feed a family and guests for days. Order crayfish ("craw-dads") at the right season and your table will be piled high with them. All of this, and the livelihoods that depend on it, is now lost over large areas.

Because of how tightly connected our economy and society are, it is not hard to foresee many of the consequences. As the owners of boats, restaurants, and motels lose business, they lay off employees and pay less tax. The suppliers with whom they do business suffer the same in connections that extend across the country and across the oceans. Oil on a beach means local restaurants serve less beef from Kansas, fewer chickens from Arkansas, and fewer vegetables from California. Those restaurants order fewer serving plates from overseas, and motels order fewer sheets, and less detergent to wash the sheets. Employees laid off will not be buying new cars or wide-screen televisions, eating out, or replacing a washing machine. Church donations are already down. With reduced taxes, state and local governments will hire fewer teachers or police officers. Such connections could be traced on and on. BP, the company that leased the Deepwater Horizon, has stated that it will pay

J.A. Tainter and T.W. Patzek, *Drilling Down: The Gulf Oil Debacle and Our Energy Dilemma*, DOI 10.1007/978-1-4419-7677-2_1,
© Springer Science+Business Media, LLC 2012

all claims, but there are limits to that commitment. Can a vegetable grower in California expect compensation for fewer shipments to Gulf Coast restaurants? What about a vegetable farmer in Mexico, or a fruit grower in Chile? Can a seafood restaurant in Albuquerque or Denver expect compensation because shrimp and oysters from the Gulf are scarcer and more expensive? For that matter, what about companies such as Zatarain's, which produces spices for New Orleans cuisine, or Café du Monde, a local coffee shop and producer of a special coffee blend? At some point people, businesses, and governments hurt by the spill will have to absorb their losses.

There are also losses that cannot be counted in money, and these may be far more tragic. A few years ago, a colleague in economics, George Peterson of the U.S. Forest Service's Rocky Mountain Research Station, was asked to help determine compensation for damages from the *Exxon Valdez* oil spill in southeast Alaska. There, as in the Gulf, people accustomed to a life of fishing suddenly lost their livelihoods. The surprise was to discover that these people could not be adequately compensated with any amount of money. People had lost a way of life that gave meaning and value. How do you compensate people who have lost their sense of worth, their identity? Quite simply, you cannot. As it was in the Alaska spill, so it is in the Gulf. Money may be necessary, but it cannot compensate for what has been lost. And knowing the people of the Gulf, we are certain that they do not want to spend years living off payments from BP.

Then there is the natural ecosystem itself, the marshes and beaches, the fish, birds, and mammals, and the once-blue water. Beaches can be cleaned, but you cannot restore a complex system. Nature must do that, and will, but the process may take decades. This is the most important restoration of all. All else in the Gulf – businesses, jobs, taxes, church donations, a way of life – depend on this natural system.

A great deal has been written about the Gulf spill in articles, books, and online. Much of this, however, repeats the obvious observations about our dependence on oil, energy independence, the desirability of clean energy, and the failures of regulation. Although we do not downplay the importance of these matters, such points are already known. Within the Gulf tragedy there are deeper lessons about energy, about our society, about how we came to be both so complex and so dependent on fossil fuels, and about what this means for our future. It is clear that the Gulf tragedy and its aftermath constitute a period in time when important lessons can be drawn and learned, and a moment when we will be open to introspection about oil and a society that requires such great quantities of this nonrenewable resource. The late anthropologist Leslie White once noted that a bomber flying over Europe during

World War II consumed more energy in a single flight than had been consumed by all the people of Europe during the Paleolithic, or Old Stone Age, who existed entirely by hunting and gathering wild foods. White estimated that such societies could produce only about 1/20 horsepower per person, an amount that today would not suffice for even a fleeting moment of industrial life. Our societies today need such vast amounts of energy that we provide it by mining stocks of solar energy accumulated eons ago, and converted into coal, natural gas, and petroleum. Without these stocks we could not live as we do.

Is it realistic to think that we can simply rely forever on today's energy sources? Groups such as the Association for the Study of Peak Oil and Gas (ASPO) warn that we will soon reach a point known as "peak oil." When this point is reached, oil production cannot be increased, even when there is plentiful oil in the ground. In fact, once production starts to decline, each year thereafter the world will need to get by on less oil than the year before. The date of reaching peak oil is controversial. The U.S. Army once predicted that it would be 2005, and some analysts – including one of the authors – think that indeed we reached it then. If so, the effects have been masked by the current recession and the development of previously unreachable oil deposits, such as in deep water. The simple answer is that we do not know exactly if peak oil has been reached, nor how long global oil production would hover at a level close to the peak. The only certainty is that the global peak of oil production is closer each day.

The Roman poet Juvenal wrote that "a good person is as rare as a Black Swan." Until 1697, when black swans were found in Australia, they were thought not to exist. All swans observed by Europeans had been white. The term has come to mean something that has never been observed, and is considered either impossible or highly unlikely. As explained by Nassim Nicholas Taleb, nothing in the past convinces us that a black swan can exist. Was the Gulf spill a Black Swan, something that was highly unlikely to happen? Nothing like it had occurred in America's waters, not even the *Exxon Valdez* spill. Most people, and clearly the regulatory authorities, thought that such a catastrophe could not happen. Yet there have actually been several times when we averted such spills because the blowout preventer, which failed in the Deepwater Horizon case, did work. Such events point to a systemic problem, and suggest that the spill was in fact likely given sufficient opportunities and time.

There is, however, still a sense in which the Black Swan metaphor is useful here. One important aspect of Black Swan events is that they give us an opportunity to see the world in a new light, to discard outdated assumptions and question what we have thought. Our society rarely thinks about our energy supply, or how that supply brings food to our tables, clothing and consumer goods to stores, loans for cars and houses, and taxes for the government. Even donations to churches depend ultimately on petroleum. Our ignorance of energy has been like the one-time ignorance of Europeans about swans. Economists treat energy as a commodity, no different from bananas or iPods, to be produced and sold in relation to market demand. Peak oil, and the resulting imperative to drill deeper and more remotely to find new oil, not only gives us the opportunity to look at the assumptions in our lives, but also the larger societal processes that result from what we call the energy–complexity spiral.

Toward the end of World War II, Vannevar Bush, director of the wartime Office of Scientific Research and Development, submitted a report to President Truman entitled *Science, the Endless Frontier*. President Roosevelt had requested the report because of the great contribution of science to the war effort. In the report, Bush wrote that

> Advances in science will…bring higher standards of living, will lead to the prevention or cure of diseases, will promote conservation of our limited national resources, and will assure means of defense against aggression.

Nearly 65 years later, Secretary of Energy Steven Chu voiced nearly the same optimism. "Scientific and technological discovery and innovation," he testified before Congress in 2009, "are the major engines of increasing productivity and are indispensable to ensuring economic growth, job creation, and rising incomes for American families in the technologically driven twenty-first century." Both statements reflect an enduring facet of American life: our optimism that technology will solve today's problems and provide a better future. The Deepwater Horizon was an expression of that optimism and, until it exploded and sank, it might have given comfort that the optimism was warranted.

Yet the Deepwater Horizon shows both the strengths and the weaknesses of our reliance on technology. Humans have been using petroleum products for 5,000 years, and in that time we have exploited the most accessible sources, those easiest to find and bring into production. As we exhaust the easiest sources, we turn to deposits that are less accessible and costlier to obtain. In the early 1930s, the Texas Co., later Texaco (now Chevron) developed the first mobile steel barges for drilling in the brackish coastal areas of the Gulf of Mexico. In 1937, Pure Oil (now Chevron) and its partner Superior Oil (now ExxonMobil) used a fixed platform to develop a field in 14 feet of water one mile offshore of Cameron Parish, Louisiana. In 1946,

Magnolia Petroleum (now ExxonMobil) drilled at a site 18 miles off the coast, erecting a platform in 18 feet of water off Saint Mary Parish, Louisiana. The Macondo Well was the technological descendent of these, and many other, early offshore wells.

There is a systematic pattern that links our demand for oil to the complexity of the technology we use to find and produce petroleum, our economic and energy return on energy production, the complexity of our society, and our ability to maintain the way of life to which we are accustomed. It takes energy to get energy, to find, extract, refine, and distribute it. The difference between what we spend and what we get back is called net energy. It is also known at Energy Returned on Energy Invested (EROEI), a term that will be even more prominent in the future. As the petroleum we extract comes from reserves that are more and more inaccessible – a mile underwater in the case of the Deepwater Horizon – the net energy declines. While EROEI declines, the technology that we develop to find and extract petroleum grows increasingly complex and costly. It takes more energy to get energy, and to develop and run petroleum technology. Deep-sea exploration rigs are among the most complex technologies that we have developed, and they are correspondingly costly. The Deepwater Horizon cost about $1,000,000,000 to build in 2001, and $500,000 a day to operate. For the past 100 years, abundant and inexpensive energy has fueled tremendous growth in the size and complexity of our societies, and in the numbers of people that the earth supports. This energy–complexity spiral means that we need greater amounts of energy just to stay even, let alone continue to grow. At the same time, our way of life and the ordinary challenges of living generate problems that require additional complexity and energy to solve. This added complexity is not just in the technological sphere, but also in our institutions, our activities, and our daily lives. The energy–complexity spiral occurs because abundant energy stimulates and requires more complexity, and complexity in turn requires still more energy.

Over the last few centuries, this spiral has moved ever upward. The question we must confront is: how much longer will this pattern continue? The spiral moves upward today in the face of greater and greater resistance, that resistance being the increasing difficulty of getting oil. The Gulf disaster forces us to confront this dilemma. It makes us see how costly it can be to pursue petroleum that is ever more remote, and to ask whether we can plan on a future that requires still more oil. The tragedy in the Gulf shows that although we need oil for our way of life, oil can also ruin that way of life directly or through our inability to manage the growing risks associated with complexity in all areas from technology to business operations to government oversight. In undertaking to write this book, then, our purposes are twofold: first to explain the Gulf disaster, the energy–complexity spiral, and how they are necessarily connected; and second to encourage all consumers of energy to consider whether this spiral is sustainable, and what it will mean for us if it is not.

Chapter 2

The Significance of Oil in the Gulf of Mexico

It was 9:15 p.m. on April 20, 2010, and the captain of the Deepwater Horizon was entertaining heavyweights from British Petroleum (BP) and Transocean, by showing off the computers and software at his disposal. After the Captain welcomed his visitors on the bridge, Yancy Keplinger, one of two dynamic-positioning officers, began a tour while the second officer, Andrea Fleytas, was at the desk station. The officers explained how the rig's thrusters kept the Deepwater Horizon in place above the well, showed off the radars and current meters, and offered to let the visitors try their hands at the rig's dynamic-positioning video simulator. One of the visitors, a man named Winslow, watched as the crew programmed-in 70-knot winds and 30-foot seas, and hypothetically put two of the rig's six thrusters out of commission. Then they set the simulator to manual mode and let another visitor work the hand controls to maintain the rig's location. While Keplinger was advising about how much thrust to use, Winslow decided to grab a quick cup of coffee and a smoke. He walked down to the rig's smoking area, poured some coffee, and lit his cigarette.[1]

Most readers will be familiar and comfortable with this narrative. There was nothing extraordinary about it, as thousands of similar scenes of human–computer and human–machinery interactions play out every day in industrial, medical, military, banking, security, or TV news settings. Everything seemed to be under control, with the computers in charge, and their sensors humming. The people assigned to watch these computers, and

---

[1] All of these events are documented in the President's Commission Report, Chap. I, p. 7.

J.A. Tainter and T.W. Patzek, *Drilling Down: The Gulf Oil Debacle and Our Energy Dilemma*, DOI 10.1007/978-1-4419-7677-2_2, © Springer Science+Business Media, LLC 2012

act on their advice, were content and getting ready to go to sleep. This is who we have become, and this is the environment in which most of us exist.

Suddenly, all hell broke loose, and it became clear that the people watching the computer screens did not understand what the computers were telling them. It took just a few seconds for their false sense of security to go up in the same flames that consumed the Deepwater Horizon in two days.

Although the outcome was extraordinary, the circumstances were not. Thousands of computer screens and messages are misinterpreted or misunderstood every day, but only occasionally does a mine cave in, a nuclear reactor melt down, a well blow out, a plane crash, a refinery explode, or soldiers die from friendly fire as a result. Each time we are reassured that the incidents were isolated and could have been avoided if people were just more thoughtful, better trained, or better supervised, managed, and regulated. Is this sense of security justified, a sort of divide-and-conquer mentality where isolated events appear small and amenable to familiar solutions, or are these events the result of societal processes over which we have little control?

Why would a company like BP build such a monument to technology and ingenuity as the Macondo well in the first place? Why was it necessary to drill for oil one mile beneath the surface of the Gulf of Mexico? Hubris among top management may have minimized the perception of risk, but well-informed employees throughout the organization understood the perils as well as the benefits of deep offshore operations. You may think that the need and motivation for these operations are obvious, but any rationale for drilling in these inhospitable environments must take into account the amount of oil (or energy in some form) that is needed to build and maintain an offshore drilling rig such as the Deepwater Horizon, extract the oil, and transport, store, and bring the precious liquid to market. In other words, large offshore platforms are built and operated using vast quantities of energy in order to find and recover even more. The cost is still higher when you consider the complex management and regulatory structures needed to complement the technology, however poorly you may feel that the responsible people performed in the case of the Deepwater Horizon.

Let us begin with fundamentals. First we need to know how much recoverable oil is waiting for us down there, how this amount of oil measures up against demand and total oil use in the United States, and how big the energy profit is after so much energy is expended in exploration, drilling, recovery, refining, and transportation to your local gas station or power plant. In other words, do the benefits outweigh the risks, for whom, and for how long?

We also need to know something about energy itself. Everybody talks about energy, but do we really understand its omnipresent role in society?

How does our insatiable appetite for energy fuel the growth of technological and organizational complexity, with all of their attendant benefits and costs? In this book, you will learn the technological and organizational factors that led to the disastrous oil spill in the Gulf. We call these factors the proximate causes. You will also see how energy and complexity can enter a positive feedback loop and spiral out of control in human societies, which makes catastrophes on local and regional scales increasingly likely, and can even threaten the future of our civilization.

How Important Is Oil Production in the Gulf?

Oil production in the Gulf of Mexico is considered vital to meeting U.S. energy needs, and thus world energy requirements. We present here some data on the Gulf oil reservoirs that we know about, those we expect are yet to be discovered, and those that we think exist but will always be too small to exploit because the potential economic or energy profit is too small. Oil or gas reserves are the quantities of crude oil or natural gas (total hydrocarbons) we are sure can be recovered profitably from known accumulations of hydrocarbons. The concept of reserves implies that oil companies can use "off-the-shelf" technology to get at the hydrocarbons. In other words, to count as reserves, the hydrocarbons must be discovered, commercially recoverable, and still remaining. Usually, only 1/3–1/2 of the oil and 3/4 of the gas in place can be recovered economically.

To estimate oil and gas reserves in the Gulf (see Fig. 2.1), we first have to define the physical extent of the oil-producing areas in what is known as the Outer Continental Shelf (OCS). In the U.S. Interior Department's lingo, OCS consists of the submerged lands, subsoil, and seabed lying between the seaward extent of the states' jurisdiction and the seaward extent of federal jurisdiction. The continental shelf is the gently sloping undersea plain between a continent and the deep ocean. The U.S. OCS has been divided into four leasing regions, one of which is the Gulf of Mexico (GOM) OCS Region.

In 1953, Congress designated the Secretary of the Interior to administer mineral exploration and development of the entire OCS through the Outer Continental Shelf Lands Act (OCSLA). The OCSLA was amended in 1978 directing the secretary to:

- Conserve the Nation's natural resources.
- Develop natural gas and oil reserves in an orderly and timely manner.
- Meet the energy needs of the country.

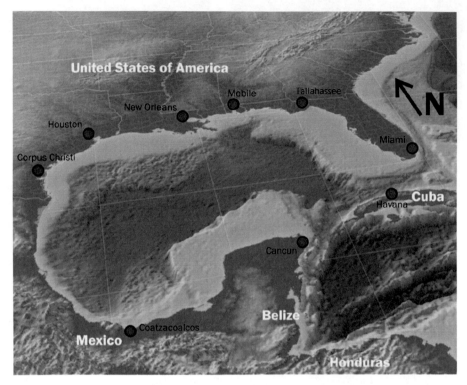

Fig. 2.1 The continental shelf of the Gulf of Mexico is topographically diverse, and includes slopes, escarpments, knolls, basins, and submarine canyons. Ocean water enters from the Yucatan channel and exits from the straits of Florida, creating the loop current associated with the upwelling and the high level of nutrient flow of this large marine ecosystem. Large quantities of freshwater are delivered from rivers in the United States and Mexico. The Gulf of Mexico is North America's most productive sea. Its shallow waters, especially river estuaries, teem with marine life. The region of the Mississippi River outflow sustains the highest level of marine life in the Gulf of Mexico. Chemical water pollution, coastal erosion, and overfishing are major threats to the health of this most important marine ecosystem in North and Central America. The Gulf of Mexico region is also a major oil and gas province that delivered 1.5 million barrels of oil per day for the United States in 2009. (Sources: NOAA, Minerals Management Service (MMS))

- Protect the human, marine, and coastal environments.
- Receive a fair and equitable return on the resources of the OCS.

  State jurisdiction is defined as follows.

- Texas and the Gulf coast of Florida are extended three marine leagues (approximately nine nautical miles) seaward from the baseline from which the breadth of the territorial sea is measured.

- Louisiana is extended three imperial nautical miles (imperial nautical mile = 6,080.2 feet) seaward of the baseline from which the breadth of the territorial sea is measured.
- All other states' seaward limits are extended three nautical miles (approximately 3.3 statute miles) seaward of the baseline from which the breadth of the territorial seaward is measured.

As you can see, Texas got a much better deal than all other states, but Texas is bigger and – some people think – better. For our purposes, suffice it to say that federal jurisdiction is defined under accepted principles of international law. Thus, the GOM OCS covers an area of over 600,000 square kilometers, a little less than the area of Texas and twice the size of Poland.

As Figs. 2.2 and 2.3 show, most of the large oil and gas fields were discovered more than 30 years ago and the future "reserve" growth will have little effect on the ultimate hydrocarbon recovery from the Gulf's OCS. The sizes of reservoirs are important for understanding ultimate oil recovery from the GOM. It turns out[2] that over the entire range of reservoir sizes, hydrocarbon reservoirs follow a "parabolic-fractal" law that says there is an increasing proportion of the smaller reservoirs relative to the larger ones. In other words, the reservoir size drops off faster than a simple power law would predict. Leaving aside the mathematics of fractals, if this law of reservoir sizes holds true, our current estimate of ultimate oil recovery in the Gulf might prove to be highly accurate, because most, if not all, of the largest oilfields have already been discovered, and the smaller ones will not add much new oil to the total regardless of how many new oilfields are discovered. On the other hand, the probability of finding another very large reservoir (a new "king," "viceroy," or at least an "elephant") is much higher than a normal or "Gaussian" probability distribution would predict. We can refer to this possibility as "fractal optimism."

Finding new oil in the deep Gulf of Mexico has not been easy. Historically, "dry holes," wells that never produced commercial hydrocarbons, have been numerous. In water depths greater than 1,000 feet (305 meters), 1,677 dry hole wells were drilled, with 331 dry hole wells in water depth greater than 5,000 feet (1,520 meters). To put the last number in perspective, 72% of all wells drilled in water depths greater than 5,000 feet were dry holes! The BP Macondo well was an exploration well that definitely was a success of sorts.

---

[2] Jean Laherrère, Distribution of field sizes in a petroleum system: Parabolic-fractal, lognormal, or stretched exponential?, *Marine and Petroleum Geology*, **17** (2000), 539–546.

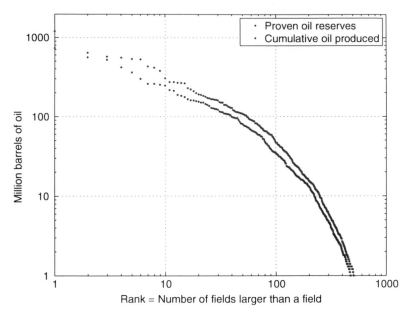

Fig. 2.2 This is the complete ranking of oil deposit volumes in the Gulf of Mexico reported to the Minerals Management Service by 2006, the latest complete statistic. The cutoff for production is one million barrels of cumulative oil produced. Thus the nonproducing oil reservoirs are excluded from the lower curve. The upper curve ranks the "proven oil reserves," (the oil we can produce for sure) with a cutoff of one million barrels of oil as well. The upper curve has 32 more points (oil fields) than the lower one, and the same ranks do not correspond to the same reservoirs. The plot has the logarithmic x-and y-axes. A simple power law, Rank×Volume^a = Constant, would be "fractal" and plot as a straight line of log Volume versus log Rank, just like this plot. The fact that both curves bend down means that reservoir size decreases faster than a simple fractal would predict. Such a distribution is a "parabolic-fractal" or a "stretched exponential." Note that the reservoir volumes do not follow a bell curve, and their distribution is not Gaussian. Mother Nature operates very differently from finance and statistics that use the Gaussian distributions *ad nauseam*, whether they are justified or not. The largest reservoirs are discovered and produced first, therefore adding new discoveries of small reservoirs is unlikely to change significantly how much oil will be ultimately produced from the Gulf of Mexico. Since 2006, however, there have been several major new discoveries by Shell and others. It is hoped these discoveries will add to the reservoir volume in the largest fields, where it counts the most

Since 1995, the overall fraction of dry holes in the Gulf of Mexico was close to 25% of all wells drilled.

The U.S. federal government has kept records of oil and gas production in the Gulf of Mexico since 1947. According to the Minerals Management Service, between January 1947 and September 2010, 46,221 wells were

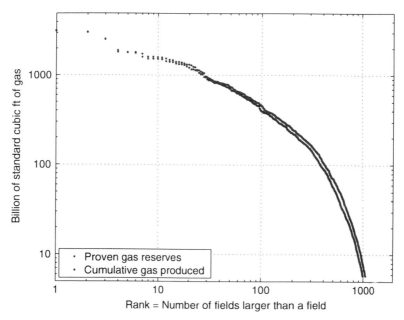

Fig. 2.3 This is the complete ranking of gas deposit volumes in the Gulf of Mexico reported to MMS by 2006, the latest complete statistic. The cutoff for production is 5.8 billion standard cubic feet of cumulative gas produced. Upon combustion, this volume of gas generates the same heat as roughly one million barrels of oil (one barrel of oil is energy-equivalent to 5,800 standard cubic feet of natural gas). The nonproducing gas reservoirs are excluded from the *lower curve*. The *upper curve* ranks the "proven gas reserves," also with a cutoff of 5.8 billion standard cubic feet of gas, equivalent in energy to one million barrels of oil. There are 62 more points on the *upper curve* than on the *lower* one, the same ranks do not correspond to the same reservoirs, and the seeming coincidence of the two curves is an optical illusion. Note that with the same lower cutoff, there are twice as many gas deposits as oil deposits, reflecting the dominance of natural gas in the Gulf. Also note the rapid proliferation of the ever smaller gas reservoirs (*the curves bend down very steeply*)

drilled in shallow Gulf water at depths of up to 1,000 feet (305 meters), and 19,888 wells are still producing. Some 3,500 platforms were activated in the shallow GOM. Between January 1975 and September 2010, 3,757 wells were drilled in deep GOM, and 1,077 wells are still producing in water depths greater than 1,000 feet (305 meters). Forty-seven platforms were activated in the deep Gulf. In water depths greater than 5,000 feet (1,524 meters), 645 wells were drilled and 115 are still producing from ten platforms. Thus, over the last 60 years, some 60,000 wells were drilled in the GOM and produced from 3,550 platforms, which is a gigantic investment

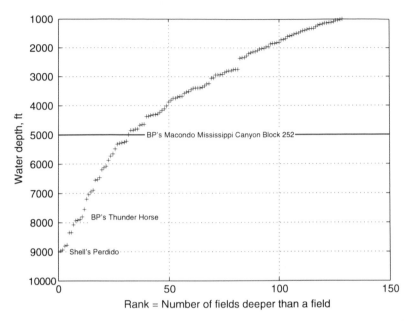

Fig. 2.4 The majority of oil production in the Gulf of Mexico comes from platforms in water deeper than 1,000 feet. There were 129 oil and gas deposits (*reservoirs*) reported by the Minerals Management Service in 2006 in water depths greater that 1,000 feet, 29 of them in water depths greater than 5,000 feet. Note that the water depth of BP's Macondo well is really 5,067 − 75 = 4,992 feet below the water surface. Its depth was measured from the derrick floor of the Deepwater Horizon rig, 75 feet above the sea. Some 1,073 wells are producing in water depths greater than 1,000 feet, 115 of them in water depths greater than 5,000 feet

of material and human resources. Figure 2.4 summarizes the distribution of known oil and gas deposits in the Gulf of Mexico in water depths greater than 1,000 feet.

The rates of oil production from the shallow (less than 1,000 feet deep) and deep (above 1,000 feet of depth, and mostly above 4,000 feet) Gulf water are shown in Fig. 2.5. The shallow water production peaked in 1973, and the deepwater production might have peaked in 2009. Our forecast is based solely on the historical production and its future decline; when completely new oilfields are brought online, our estimate may go up. The cumulative oil produced from the deepwater Gulf is shown in Fig. 2.6. The industry forecasts up to nine billion barrels of ultimate production from the deepwater Gulf, whereas Patzek forecasts only eight billion barrels. Either way, the total oil produced from the deep Gulf water will be less than 11 billion barrels of oil already produced from the Prudhoe Bay field in Alaska, with another

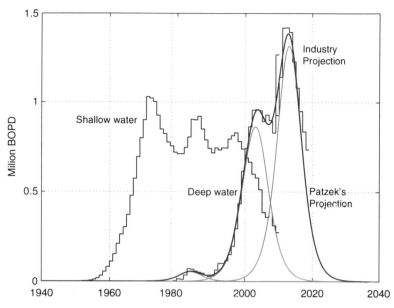

Fig. 2.5 The rate of oil production from the offshore Gulf of Mexico in millions of barrels of oil per day (BOPD). In 2010, the United States used about 19 million BOPD, and produced five million BOPD. Note that the shallow water development occurred in three stages. The highest peak of shallow oil production occurred in 1971, and the production rate has been declining rapidly since 1997. Deepwater oil production became noticeable in 1979, reached the first peak in 2003, and jumped to a new all-time high in 2009, mostly because of BP's Thunder Horse. The industry forecasts the new large oilfields coming online soon, and peaking in 2012–2013. Because we have no knowledge of the production capacity of the new finds in the GOM (these finds are closely held secrets), we project a faster decline based on the Hubbert curves that fit historical production data and predict the future decline of the oilfields in the dataset. The right Hubbert curve is related mostly to ultra-deepwater. This curve will probably grow as the future oilfields start producing. (Sources: U.S. DOE Energy Information Administration (EIA), Minerals Management Service (MMS), and Patzek's calculations)

one billion barrels to go at Prudhoe. The Gulf oil production will also be roughly one third of the oil produced by Norway in the North Sea. There is only one Prudhoe Bay in North America and there are some 400 producing oil reservoirs in the U.S. Gulf, with an estimated 900 small reservoirs yet to be produced. Such is the fundamental injustice of Mother Nature: one supergiant oil field can produce more oil than a large geographical region of a continent, the Gulf Coast. To add insult to injury, it is also much cheaper to produce oil from Prudhoe Bay than from deepwater.

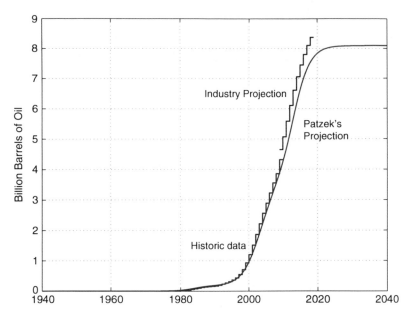

Fig. 2.6 Cumulative oil production from the deep offshore Gulf of Mexico in billions of barrels of oil. Based on the historical data, we estimate that the deep Gulf will produce about eight billion barrels, whereas the industry projects about nine billion barrels. We will adjust our prediction upward as the new production data become available. Even nine billion barrels of oil is less than the 12 billion barrels of ultimate oil production from Prudhoe Bay in Alaska, the largest oilfield in North America. Thus far, Prudhoe Bay has produced 10.8 billion BO. Ultimately, the deep GOM will produce about a third of the oil produced by Norway in the North Sea. (Sources: U.S. DOE EIA, MMS, and Patzek's calculations)

An average well in the GOM is not very productive. Gas wells account for 60–70% of new wells, and only 30% are oil wells. The mean gas production rate is 1.5 million standard cubic feet per well per day (CFPD) with the current maximum of over 100 million CFPD. The mean oil production rate is 450 stock tank barrels of oil per day (STBOPD) per well, with the current maximum of 41,000 STBOPD. Since January 2008, only 10% of GOM wells have been producing in excess of 2,000 STBOPD or 11 million CFPD.

Nevertheless, in inflation-adjusted dollars (see Fig. 2.7), the total revenue produced from the Gulf has been a little less than $700 billion. Seven hundred billion dollars is as much as the Troubled Asset Relief Program (TARP) that allowed the U.S. Department of the Treasury to purchase or insure up to

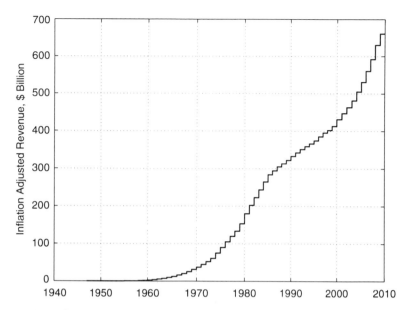

Fig. 2.7 Cumulative gross revenue generated by the offshore Gulf of Mexico oil. The average oil prices have been adjusted for inflation to June 2010 prices using the Consumer Price Index (CPI-U) as presented by the Bureau of Labor Statistics. (Sources: U.S. DOE EIA, MMS, BLS, and Patzek's calculations)

$700 billion of "troubled assets[3]". It took the hard work of three generations and tens of thousands of people to produce this *real* oil wealth in the Gulf over 70 years, but an equal amount of wealth was annihilated by a few hundred speculators, peddling bets rather than real assets, over just a few years.

Figure 2.8 illustrates the bottom line. The ratio of oil production from the Gulf to production from all other oilfields in the United States outside of the Gulf was increasing until 2003, when it was equal to 37%. Since then this ratio decreased to about 30% in 2008, and jumped back to 40% in 2009, with oil production elsewhere in the United States decreasing slowly.

---

[3] Defined as "(A) residential or commercial mortgages and any securities, obligations, or other instruments that are based on or related to such mortgages, that in each case was originated or issued on or before March 14, 2008, the purchase of which the Secretary determines promotes financial market stability; and (B) any other financial instrument that the Secretary, after consultation with the Chairman of the Board of Governors of the Federal Reserve System, determines the purchase of which is necessary to promote financial market stability, but only upon transmittal of such determination, in writing, to the appropriate committees of Congress." Source: *CBO Report, The Troubled Asset Relief Program: Report on Transactions Through December* 31, 2008.

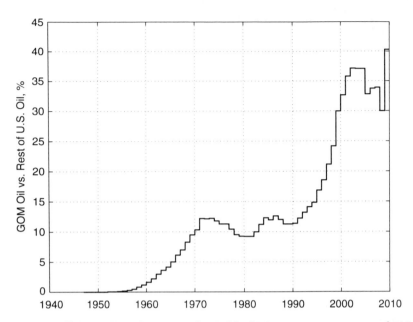

Fig. 2.8 All offshore oil production in the Gulf of Mexico as a percentage of U.S. oil production from all sources outside of the GOM. Note that the ratio of the GOM's oil production to the remaining U.S. production first peaked in 2003, at 37%, and then declined until 2009, when it jumped to 40%, in good part because of BP's Thunder Horse. (Sources: U.S. DOE EIA and MMS)

The total production in the United States has hovered around five million barrels of oil per day since 2004, and it may increase again in 2011 because of all the new drilling in Texas and North Dakota. The new production from BP's very large Thunder Horse semisubmersible platform[4] resulted in a jump of 260,000 barrels of oil per day, and the actual production in the GOM is a little ahead of the industry projection.

---

[4]Thunder Horse production and drilling quarters is the world's largest production semisubmersible platform ever built. The platform's topside area is the size of three football fields. It is packed with equipment and systems capable of processing and exporting a quarter of a million barrels of oil per day. Thunder Horse never hit that level of production. This semisubmersible produces oil and gas in one of the largest hydrocarbon fields in the Gulf of Mexico. The floating platform operates under extreme conditions. It pumps out oil and gas from the field which is three miles beneath mud, rock, and salt, topped by a mile of ocean. The hydrocarbon pressure is over 1,200 atmospheres (17,600 psi) and its temperature is 135°C (275°F). More information can be found at www.offshore-technology.com/projects/crazy_horse/ and www.petro-leumnews.com/pntruncate/10942334.shtml.

Because of the huge flash production from Thunder Horse, and the need for more U.S. oil, the Interior Department exempted BP's plans for a nearby Macondo well from a detailed environmental impact analysis in 2009. According to government documents, after three reviews of the area it was concluded that a massive oil spill was unlikely.[5] On April 6, 2009, a fateful decision was handed down by the department's Minerals Management Service (MMS) to give BP's lease at the Mississippi Canyon Block 252 a "categorical exclusion" from the National Environmental Policy Act (NEPA). Neither the government nor BP conceived of the possibility of a Black Swan, in Taleb's memorable term for an unpredictable or unlikely event, descending upon the United States only 1 year later, almost to the day. Taleb's book, *The Black Swan – The Impact of the Highly Improbable*, should be required reading for all governmental and corporate managers.

Lessons About Technology and Oil Reserves

Unfortunately, oil production from the prolific Thunder Horse has been declining faster than most experts thought. In the words[6] of Glenn Morton, a consultant for oil exploration projects and analyst for OilDrum.com, "Thunder Horse hasn't reached anywhere near its expected potential," in oil or natural gas, a fact which "underscore[s] the point that deepwater oil drilling is a tricky process, and not always as easy or predictable as thought." Thus Lesson Number 1: the complex technology we deploy to conquer Nature interacts with the complex Earth systems in ways that are either unpredictable or very difficult to quantify.

In 2005, during Hurricane Dennis, an incorrectly plumbed, 6-inch. length of pipe allowed water to flow freely among several ballast tanks that ultimately caused the Thunder Horse platform to tip into the water; see Fig. 2.9. The platform was fully righted about a week after the hurricane, delaying commercial production initially scheduled for late 2005 by 3 years. During repairs, it was discovered that poorly welded pipes in an underwater manifold were severely cracked, and the $1 billion manifold had to be redone. Lesson Number 2: the small cheap parts of the astronomically expensive complex

---

[5] "U.S. Exempted BP's Gulf of Mexico Drilling from Environmental Impact Study," Juliet Eilperin, *Washington Post* staff writer, Wednesday, May 5, 2010.
[6] See www.energybulletin.net/node/52659.

Fig. 2.9 The U.S. Coast Guard reported on July 12, 2005: "Thunder Horse, a semi-submersible platform owned by BP, was found listing after the crew returned. The rig was evacuated for Hurricane Dennis." This almost $1 billion platform was nearly sunk by an incorrectly plumbed 6-inch. pipe valued at just a few dollars. Then there were problems with welds in the subsea pipes and manifolds. Their replacement cost and 3 years of production delays might have been as high as another billion USD. (Image Source: USCG photo by PA3 Robert M. Reed, displayed in Wikipedia)

structures can cause them to disintegrate. Remember the failed rubber O-ring in the right booster rocket of the space shuttle Challenger? We show how this lesson was missed at the failed BP Macondo well.

We must also not forget that drilling for offshore oil has always been hard and dangerous work. Since 2001, the Gulf of Mexico workforce – 35,000 people, working on 90 big drilling rigs and 3,500 production platforms – has suffered 1,550 injuries, 60 deaths, and 948 fires and explosions.[7] Almost

---

[7] Bureau of Ocean Energy Management, Regulation, and Enforcement, Installations, Removals, and Cumulative Totals of Offshore Production Facilities in Federal Waters; 1959–2010, 2/2010, www.boemre.gov/stats/PDFs/OCSPlatformActivity.pdf.; Bureau of Ocean Energy Management, Regulation and Enforcement, OCS Incidents/ Spills by Category: 1996–2005, 10/19/2007, www.boemre.gov/incidents/ Incidents1996–2005.htm; Bureau of Ocean Energy Management, Regulation, and Enforcement, OCS Incidents/Spills by Category: 2006–2010, 7/10/2010, www.boemre.gov/incidents/IncidentStatisticsSummaries.htm.

20% of all fatalities in the GOM were caused by the BP Macondo well blowout.[8]

The Take-Home Messages

The cover photo of this book is familiar to anyone who watched television or internet news sources or read a newspaper in April, 2010, and should be sufficient to convince most people that offshore production of oil and gas is risky. Because the Gulf of Mexico currently accounts for 40% of crude oil produced in all other areas of the United States combined, drilling in the Gulf is clearly necessary in spite of the risks and the diminishing returns on investment. Ultimate oil recovery from OCS will likely exceed nine billion barrels of oil, roughly 1.5 years of the total U.S. consumption of crude oil and petroleum products in 2009. In other words, the deep and ultra-deepwater oil production in the GOM will suffice to power the United States for 1.5 years.

Most oil production in the GOM comes from water depths greater than 1,000 feet, and a good portion from depths below 4,000 feet. Although Gulf of Mexico oil accounted for about 8% of U.S. daily oil use in 2010, and this percentage is likely to decline in the next few years, stopping GOM production would have severe economic repercussions for the U.S. economy.

Exploration in the deep Gulf is financially risky, with over 70% of all new wells never producing appreciable quantities of hydrocarbons. Given the price tags of these wells, anything from $50 to $200 million apiece, deepwater drilling in the Gulf is not for the faint-hearted. After some of the current very large finds by Shell, BP, and others are produced, oil production in the deep Gulf will move to the shallower and much smaller reservoirs that have been bypassed on the way to the deep big ones. Elsewhere in the world, oil exploration and production will go deeper, even much deeper, in hopes of larger rewards. Oil production in the Gulf will continue for several more decades, albeit at a much decreased level. Currently, there are 1,073 successful producing wells in water depths greater than 1,000 feet, and 115 in water depths greater than 5,000 feet. In recent years, about 400 new gas and oil wells have been drilled in the GOM annually.

---

[8] If the GOM workforce had fought in Afghanistan over the same period of time, one would expect at least 520 deaths, and 3,000–4,000 injuries.

In many ways, drilling for oil and gas in one mile of seawater is more unforgiving than sending people to outer space. The total darkness and remoteness, crushing pressures, near-freezing water temperature, extremely high hydrocarbon temperatures, aggressive corrosive gases, solid hydrates and paraffins, and so on, all have made ultra-deepwater perhaps the most inhospitable environment on Earth. Yet, we will continue to explore and drill in this environment, simply because there is a lot of oil and gas down there, the low-hanging fruit of easily accessible oil has been mostly picked, and our voracious appetite for concentrated sources of energy only increases with time.

The U.S. domestic oil supply peaked in 1970, and global peak oil production may have been reached in 2005 when the highest conventional oil production rate worldwide was recorded. Therefore, it is unlikely that we can obtain supplies from elsewhere to replace oil from the Gulf. If the rate of oil consumption in the United States compels oil companies and the nation as a whole to assume the risk of drilling in deep water, whose responsibility is it to manage this risk, or at least to try? The answer to this question is complex, and requires a deeper understanding of the origins of different risk factors and drivers, from the technology, to the corporate boardroom, and ultimately to societal processes that have led us to seek energy in remote inhospitable environments. Our search for an answer will take us from the Macondo well to the Roman and Byzantine empires and back. Hold on to your seats.

Further Reading

1. Mandlebrot, B.B.: Fractals and Scaling in Finance – Discontinuity, Concentration, Risk. Springer, New York (2010)
2. Taleb, N.N.: The Black Swan – The Impact of the Highly Improbable. Random House, New York (2010)
3. Zipf, G.K.: Human Behavior and the Principle of Least Effort – An Introduction to Human Ecology. Hafner, New York (1965)

Chapter 3

The Energy That Runs the World

At their roots, territorial conflicts in human and nonhuman species alike are attempts to control more of the products of solar energy conversion. More land means a greater share of incident solar radiation. In human societies, the resulting wealth can be obtained indirectly through the payment of taxes or tribute by subject peoples. This point is pertinent to our energy future, and we have much more to say about it in Chaps. 5 and 6. Solar energy can also be stored underground, and for the past 250 years humans have gained access to vast quantities of compressed and chemically transformed fossil organisms in the form of oil, coal, and natural gas. Without these concentrated stores of solar energy, which we call fossil fuels, the technological triumphs and economic progress of the past 150 years would have been the stuff of deluded imaginations.

The fuels themselves are tangible enough, but just what is energy, what forms does it take, how do we convert one form of energy into another that is more useful to us, and what if any are the negative consequences? To understand these issues, and to go beyond the wishful thinking that pervades discussions of energy supplies in general and so-called alternative and renewable energy sources in particular, we must say something about those dreaded "laws of physics." These laws tell us clearly what is possible in nature and what is not.

In 1956, a distinguished British physicist, Sir Charles Snow, greatly upset nonscientists by saying, in effect, that their lack of knowledge of the second law of thermodynamics was equivalent to scientists not ever reading a work of Shakespeare. The 1956 remark by Snow seems quite timely in 2011, when

J.A. Tainter and T.W. Patzek, *Drilling Down: The Gulf Oil Debacle and Our Energy Dilemma*, DOI 10.1007/978-1-4419-7677-2_3, © Springer Science+Business Media, LLC 2012

science is often used to justify a position already taken rather than the means to discover the truth. In this respect, "science-based" becomes the science that supports what we believe, and "junk science" is the science that disagrees with our view of the world. Energy is so essential to almost every aspect of modern life that such an attitude applied to our energy future will contribute greatly to the quick demise of our advanced society. In fact, we show in later chapters how energy supplies were a primary determinant of the survival or collapse of past civilizations.

The famous British astrophysicist, Sir Arthur Stanley Eddington, began his *The Nature of the Physical World* as follows:

> The law that entropy increases – the second law of thermodynamics – holds, I think, the supreme position among laws of nature. If someone points out to you that your pet theory of the universe is in disagreement with Maxwell's equations – then so much the worse for Maxwell's equations. If it is contradicted by observation – well, these experimentalists do bungle things sometimes. But if your theory is found against the Second Law of Thermodynamics, I can give you no hope; there is nothing for it but to collapse in deepest humiliation.

Much of the current public discourse about sources of energy, and the future of humankind is at odds with the Second Law of Thermodynamics. Later in this chapter and in Chap. 9, we consider solar and wind power as well as biofuels in the context of their contribution to U.S. consumption of energy, and what is possible to achieve and what is not in our quest for fossil fuel substitutes. What follows is a brief description of energy, the most fundamental concept in science and in society, and one that is not generally well understood, especially by the economists and bankers to whom we eagerly entrust our life savings, or by politicians to whom we offer our support in exchange for unsubstantiated claims about domestic energy supplies and promises of energy independence.

Thermodynamics

Thermodynamics is the science of admissible conversions of energy. Work, heat, and radiation are the means by which different forms of energy flow from one place to another. Work and heat are not equivalent. A quantity of work can be converted entirely to heat, as demonstrated in 1843 by James Prescott Joule, a wealthy brewer and experimental physicist. However, the young genius, Nicolas Léonard Sadi Carnot, showed in 1824 that a quantity of heat cannot be converted to an equal quantity of work in a machine that works continuously (a steam or car engine). In other words, no internal combustion engine can be 100% efficient.

All human activities require energy, but what is it? The United States wants "energy security." But what kinds of energy are there, how much do we need, and what are the available sources?

Energy, coined from the Greek word *energeia*, meaning an activity or operation, is an abstract term that denotes ability to do work, such as in the lifting of a weight or moving a car and its occupants. As explained by the historian of science and philosopher, Philip Mirowski, in his inspired book, *More Heat than Light: Economics as Social Physics*, the deeper we think about energy the more abstract and esoteric it becomes. In fact, as Mirowski shows, false translations of the concept of energy into economic terms are at the core of the irreparably flawed foundation of modern economics.

To refresh readers' memories (or nightmares) of high-school physics, the first two classical laws of thermodynamics can be approximated as follows:

- Mass and energy are conserved.
- The qualities of matter and energy deteriorate (become less useful to us) over time.

Whatever happens[1] to a chunk of matter with time, the first law above makes it clear that the quantity of each chemical element comprising this chunk does not change. For example, if you burn 12 grams of solid coal in 32 grams of gaseous oxygen, you obtain exactly 44 grams of gaseous carbon dioxide that has carbon chemically bonded with oxygen. You can convert matter from one form to something more useful (and chemists do this every day), but an amount of energy is required and the elements of matter are conserved.

The first law above also states that one type of energy converts to another one of equal quantity. For example, the kinetic energy of a car hitting a tree is converted to heat and the crumpling and tearing of the car body, but, we hope, not the occupant, into pieces (see Fig. 3.1). This law, which can be derived as a corollary to the first law of thermodynamics, implies only that energy is conserved. But we already see that another problem arises. After the crash, the mass and energy embodied in the car components are no longer in a form that we can readily use without causing increased deterioration somewhere else. For example, the scrap metal dealer can reuse the car, but then

---

[1] For simplicity, the matter can only undergo chemical, but not nuclear reactions.

Fig. 3.1 The kinetic energy of this car was entirely converted into tearing it up into pieces and heat. The car's mass was conserved. The car's kinetic energy and some of its internal energy were exchanged with the surroundings, and also conserved. But the car's elegantly manufactured structure was not. We say that the intense energy exchange between the car and its surroundings has increased entropy of both upon impact. (Image from day 24: *cell phones as dangerous as drunk driving*, Matt's blog)

much more energy is required to recycle the car's materials for new uses, and more waste products will accumulate in the environment. The problem of irreversibly losing an ordered structure is the domain of the Second Law of Thermodynamics.

The second law as stated above is really a simplified statement of the Second Law of Thermodynamics. A better statement of this aspect of the law was proposed by Professor Frank Lambert in his essay, "Shakespeare and Thermodynamics: Dam the Second Law!" Energy tends to flow from being concentrated in one place to becoming diffused and spread out, and therefore less concentrated. In this process, matter also tends to become more spread out or made random or "disorderly" if adequate energy flows through it. A hot pizza removed from the oven is the prototypical illustration of thermal energy becoming less concentrated; the heat always flows from the hot pizza to the cooler room. All types of energy tend to behave in the same fashion.

Fig. 3.2 A car was set on fire. When gasoline reacted with oxygen, the difference in internal energy between the reactants and products was spread out to the environment as intense heat and light. The gasoline, which quickly released its chemical energy when ignited by a match, would be chemically inert and intact for many years if kept in a closed cool tank. The Second Law of Thermodynamics allows gasoline to combust in principle and achieve a stable equilibrium with the atmosphere as carbon dioxide and water, but it gives no indication how soon if ever this can happen. An energy barrier to start combustion must be overcome by a spark and process kinetics govern the rate of combustion. Chemists call this barrier "activation energy." The same mechanisms acted in the methane explosion when the Deepwater Horizon ignited due to a spark from an electrical motor that did not shut off automatically. Human error presented an opportunity for the Second Law of Thermodynamics to act and destroy the ship's elaborate structure in less than 2 days. In scientific language, entropy was created. In common language, all hell broke loose

The Second Law of Thermodynamics defines what tends to or can happen, but as illustrated in Fig. 3.2, it gives no information on how quickly that something actually occurs. The second law is so special because it is the only law of physics that defines the direction of time and allows for the possibilites of something happening in the future, or not.

Here are two examples of how the second law constrains what is possible. It would take more energy and environmental resources to regenerate a chunk of coal that was just burned to carbon dioxide and heat than the amount of heat we originally obtained from that chunk. The waste products from the regeneration process would also cause greater deterioration of the environment than the original coal burning. The same principle applies to the production of corn ethanol. The energy it would take to restore the environment damaged by the corn and ethanol production processes exceeds several-fold the amount of combustion heat we get from burning the ethanol in our cars. In both examples, we cannot break even no matter how hard we try!

Considerable energy flows are required to maintain complex structures that are far from equilibrium, including living organisms and societies. We must breathe, drink, and eat for energy to flow continuously through our living bodies and maintain their highly complex, organized structures. Unfortunately, this tends to create a mess in the environment that surrounds us.

The validity of mass conservation, as well as the First and Second Laws of Thermodynamics, has been confirmed by countless reproducible experiments. If just one experiment refuted any one of these laws, all three would have to be revised. But do not hold your breath, no changes are forthcoming.

In 2007, Patzek was invited to Paris to speak at a meeting of the European ministers of environment and transportation. At the opening dinner, he sat next to a young lawyer from the U.S. Department of State. The lawyer claimed that all laws of physics have loopholes that can be used to society's benefit by the daring entrepreneur. Patzek had to break the bad news to the good lawyer: there are no loopholes in the fundamental laws of physics. If there were, scientists would have to revise these laws so that no loopholes remained. In other words, if someone's invention or theory violated one or more of the laws listed above, especially the subtleties implied by the energy dissipation under the second law, there would be no hope for the author and no limits to the professional disgrace he or she would face.[2] In contrast, social laws have many loopholes that are commonly used by lawyers to their clients' benefit.

A society is no different. Every society needs a steady flow of mass and energy (goods and services) to maintain its complex structure, and increasing flows if it increases in size or complexity. The more complex the society is, the more energy throughput it requires just to maintain its far-from-equilibrium organization. In 2008, for example, close to 1.5 months worth of all electricity generated in the United States was used to power the Internet: data servers, transmission lines and routers, and personal computers. This amount is four

---

[2] Some may remember cold fusion, which burst into notoriety on March 23, 1989, when an announcement came from Drs. Fleischmann and Pons at the University of Utah. Both were disgraced and forced out.

times more electricity than is generated in the United States from all renewable sources, including solar, wind, and biomass.[3]

In Chaps. 5 and 6, we show that the converse is also true. Size and complexity will increase in response to increased flows of energy.

Units of Energy

For the unit of energy we use 1 joule (J), named after James Prescott Joule. It is a fairly small amount of energy. A little more than 4 joules are necessary to heat one teaspoon of water by one degree on the Celsius scale. For the unit of power we use one joule per second (J/s) or 1 watt (W). From a detailed demographic calculation, it turns out that on average a U.S. resident requires 100 watt continuously to live and work. This requirement makes an American resident equivalent to one 100 watt bulb operating continuously.

Larger energy units are the powers of 1 joule. We use kilo joules (kJ), mega joules (MJ), giga joules (GJ), tera joules (TJ), peta joules (PJ), and exa joules (EJ).

Here is the list of these derived energy units.

- 1 kJ is 1,000 or 10^3 joules
- 1 MJ is 1,000,000 or 10^6 joules
- 1 GJ is 1,000,000,000 or 10^9 joules
- 1 TJ is 1,000,000,000,000 or 10^{12} joules
- 1 PJ is 1,000,000,000,000,000 or 10^{15} joules
- 1 EJ is 1,000,000,000,000,000,000 or 10^{18} joules
- During 1 year, the U.S. population requires approximately

$$100\frac{J}{s \times person} \times 300,000,000\,persons \times 3,600 \times 24 \times 365s\,/\,year \approx 1EJ\,/\,year$$

as digested food; see Fig. 3.3.

The amount of energy required to feed the entire U.S. population for 1 year, 1×10^{18} joules or 1 exa joules, is the fundamental unit in which all other energy flows in the U.S. economy are described.

---

[3]The IEA study warns that energy used by computers and consumer electronics will not only double by 2022, but increase threefold by 2030. IEA Executive Director Nobuo Tanaka said in a press release accompanying the report that the increase was equivalent to the current combined total residential electricity consumption of the United States and Japan. Source: *The New York Times*, 5/14/2009.

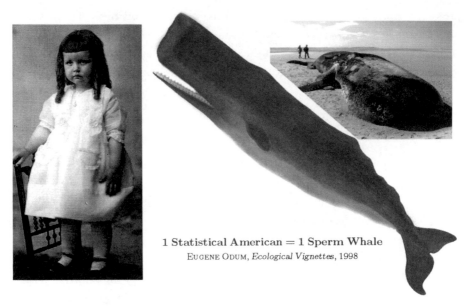

1 Statistical American = 1 Sperm Whale
EUGENE ODUM, *Ecological Vignettes*, 1998

Fig. 3.3 Our preferred unit of energy is 1 exa Joule = 1 EJ = 1,000,000,000,000,000, 000 J. This amount of energy, when metabolized as food, is sufficient to feed the entire U.S. population for 1 year. Currently the United States uses 100 EJ/year; 100 times more than we need to live. If we were to metabolize this amount of energy, we would be 15-meter-long (as tall as a five-story building) bull sperm whales, each weighing 40 tonnes. This analogy is not as far-fetched as it may seem. Recently, a similar scaling was found for power consumption per capita versus GDP per capita in 200 economies worldwide. Societies scale their "metabolism" with size identically to all mammals. There are ~300,000 sperm whales worldwide and 310 million Americans. Parenthetically, 1 EJ/year ≈ 10^{12} (tera) standard cubic feet of natural gas per year = 1 Tcf/year = not quite 5% of annual consumption of natural gas in the United States. One hundred exa Joules of heat generated in 1 year mostly from fossil fuels is equal roughly to 3×10^{12} watts of heat = 3 TW, continuously

Another unit of energy is 1 kWh (kilo watt-hour), equal to 3.6 mega joules. An average household electricity bill is about 1,000 kWh/month, or 3.6 GJ/month. To produce this electricity with an overall efficiency[4] of

---

[4]The second law of thermodynamics prevents all heat from being converted to electricity. The best coal-fired power stations in the world have efficiencies of 45–48%. The U.S. average for coal-fired power plants was 30% in 2009, down from 37% in 2002. The main reason for this dismal efficiency is the unwillingness of U.S. utilities to upgrade old and decaying power plants. This unwillingness in turn is caused by the requirements of the Clean Air Act, which forces operators to upgrade all the way to the state-of-the-art, if they decide to change anything in an old power plant.

0.32, we need to burn 11.25 GJ/month of a fossil fuel, or 135 GJ/year. This amount of energy is equivalent to the total heat from burning 2.9 metric tonnes of gasoline. So each household in the United States is responsible for burning almost 3 metric tonnes of gasoline equivalents (such as coal, natural gas, nuclear energy, hydropower, wind energy, geothermal heat, etc.) per year to obtain the needed electricity.

On average, a U.S. resident uses 100 times more energy than he needs to live, $100 \times 100 = 10,000$ watts continuously, or 0.01 mega watts (MW) of power. As an industrial worker, a single person uses and outputs many times more power. For example, it can be calculated that an average agricultural worker in the Midwest uses 0.8 MW of power as fuel for machinery, electricity, and field chemicals, and outputs 3 MW of power as crops. An average oil industry worker in California uses 2.8 MW of power as fuel for machinery and electricity, and outputs 14.5 MW of power as oil and some gas. Therefore, an agricultural worker in the United States has at her disposal the power of 8,000 ordinary people, and an average oil industry worker develops the power of 28,000 people, four times the size of an average U.S. town of 6,600 people. This external, or exosomatic, use of mostly fossil fuel power has no parallel in human history, and will be a short-lived phenomenon by evolutionary and historical standards.

How Do We Use Fossil Energy?

As described above, every man, woman, and child in the United States uses about 100 times more energy than needed to live. For many of us, the end result is an enlarged waistline.[5] Our essential reliance on fossil fuels is best illustrated by looking at the two largest sectors of the U.S. economy: electricity generation and transportation. These two sectors use about 70% of all energy consumed in the United States.

If electricity to your home goes down temporarily, you feel inconvenienced and itch to pick up your cell phone to call the utility. Other customers have similar ideas, and the utility communication networks get overloaded in no time. To keep our internet, telephones, computers, television sets, refrigerators, and lights going, we require electricity 24 hours a day and 7 days a week. It is therefore instructive to know what sources of energy provide electricity

---

[5] It is estimated that Americans dump as trash, and otherwise waste, half of the U.S. food supply.

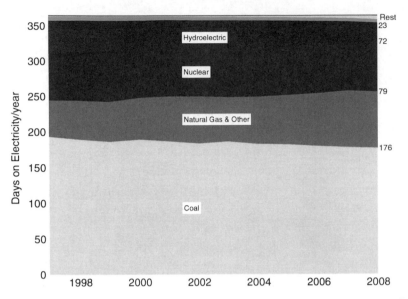

Fig. 3.4 Electricity generation in the United States uses 37% of primary energy (heat-generating energy), more than any other sector of the energy economy. In 2008, coal delivered 176 days of electricity; natural gas, 79 days; nuclear power, 72 days; hydropower, 23 days; and all remaining sources combined, 15 days. Overall, 70% of electricity in the United States was generated from purely fossil fuel sources in 2008. (Data source: DOE EIA, accessed 03/28/2010)

in our outlets 24/7/365. The data from the U.S. Department of Energy are plotted in Figs. 3.4 and 3.5 as days of electricity per year supplied to U.S. customers from each major primary energy source. It turns out that between 1997 and 2008, coal supplied between 176 and 200 days of electricity to all U.S. outlets. Coal fuels the base-load power stations and its consumption has been remarkably constant. Similarly, in 2008, natural gas supplied 79 days of electricity, nuclear power 72 days, and hydropower 23 days. The electricity share of natural gas grew from 50 days per year a decade ago, to almost 80 days in 2008. No new nuclear reactors have been built in the United States since the late 1970s,[6] but the nuclear power industry has learned how to manage their reactors better, and their share of electricity generation has grown somewhat. The share of hydropower has decreased substantially over

---

[6] According to the U.S. Dept of Energy, the last reactor built was the River Bend plant in Louisiana. Its construction began in March of 1977. The last plant to begin commercial operation was the Watts Bar plant in Tennessee, which came on line in 1996.

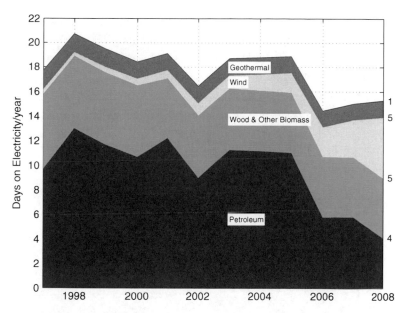

Fig. 3.5 The remaining 15 days of electricity generation in the United States can be split as follows: in 2008, 4 days of electricity came from petroleum (essentially a single power plant in Honolulu, Hawaii); 5 days from biomass burning to cogenerate electricity from wood chips, agricultural residues, and the like; 5 days from wind turbines, and growing fast, especially in Texas; 1 day from geothermal steam, mostly in California; and 1 hour from solar photovoltaics and solar thermal concentrators, also mostly in California. When someone tells you that electricity generation from photovoltaics doubles every year or so, please understand its tragically negligible significance. (Data source: DOE EIA, accessed 03/28/2010)

the last decade because of droughts and dam silting. Together, these four basic sources of primary energy delivered 350 days of electricity in 2008. The remaining 15 days of electricity were delivered by petroleum (4 days), biomass (5 days), wind turbines (5 days), geothermal steam generators (1 day), and photovoltaics/solar thermal concentrators (1 hour).

Sadly, the solar collector and photovoltaic (PV) solution is far from satisfying a meaningful fraction of our current energy needs, and will remain so for the next 20–30 years, regardless of the heroic efforts of so many outstanding researchers and companies. In fact, at the end of 2006 (the latest quality-controlled data), the total peak electrical power from the PV cells installed in the United States was 500 MW_e, or ~500/5 = 100 MW_e of continuous power,[7] equal to that of *one* small conventional electricity generator powered by coal. Usually, there is an equivalent of 10–20 such generators in a single electric power plant. Photovoltaics delivered roughly 0.006 EJ of primary energy in 2007, or 6 parts in 100,000 of primary energy needs in the United States.

The transportation sector is totally dominated by petroleum, as shown in Fig. 3.6, assurances to the contrary from the Renewable Fuels Association

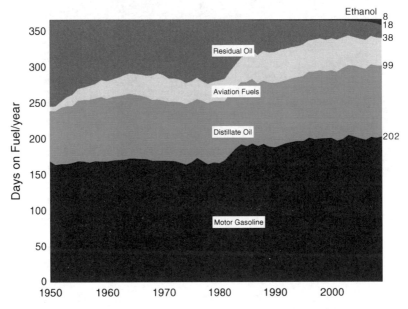

Fig. 3.6 Consumption of liquid transportation fuels in the United States accounts for 31% of primary energy (heat-generating energy), and is the second largest sector of the energy economy. In 2008, automotive gasoline delivered 202 days of transportation; distillate oil (mostly diesel fuel), 99 days; aviation jet fuel, 38 days; residual oil, 18 days (now used almost exclusively to power ships); and ethanol, 8 days. When someone tells you that driving your car on ethanol will make the United States energy-secure, laugh, because fossil fuels provide 98% of energy used to power U.S. transportation. Oil shortages caused by the peaking of global oil production will cause disproportionate disruptions in transportation, but none in electricity supply. This is the reason why a crash program in railroad electrification would benefit the United States enormously. (Data source: DOE EIA, accessed 03/28/2010)

---

[7]The EIA accounting for electricity actually produced from photovoltaics in 2007 was 70 MW_e of continuous power, www.eia.doe.gov/cneaf/solar.renewables/page/wind/wind.html, Table 1.11.

notwithstanding. Our freedom from petroleum-based liquid transportation fuels currently lasts for only 1 week per year and perhaps 2 weeks in the future. In 2008, automotive gasoline provided 202 days of all transportation needs in the United States, diesel fuel 100 days, and jet fuel 38 days. Residual or bunker oil used to be burned in electrical power stations up until 1985, but now it powers ships almost exclusively. Ethanol sufficed to power transportation in the United States for 8 days in 2008, and ethanol production itself requires a significant amount of fossil fuels.

By comparing Figs. 3.4 and 3.6, it is clear that the only way out of the total dependence on petroleum-based liquid transportation fuels is to electrify railroads, and move goods and people across the United States using electricity. Currently, there are no other choices, including biofuels and hydrogen.

Resource Versus Production

In this book we focus on liquid petroleum[8] or crude oil, a naturally occurring, flammable liquid. Petroleum is a complex soup that consists of hundreds of ingredients: hydrocarbons of various molecular weights and other liquid organic compounds that contain oxygen, sulfur, nitrogen, phosphorus, metals, and so on. Petroleum is found in massive rock layers beneath the Earth's surface and is recovered through drilling and pumping (Fig. 3.7). Subsequently, it is reformed and refined by chemical processing, and separated, most easily by boiling, into a large number of consumer products, from gasoline and diesel fuel to jet fuel, lubrication oils, asphalt, and chemical reagents used to make plastics, cosmetics, and pharmaceuticals. Petroleum permeates almost all manufactured products and is at the core of our energy-intensive civilization. Without petroleum, modern economies would not exist. Many people do not recognize this fundamental truth about their own lives, and a currently popular delusion goes something like this: because a modern society depends less on energy in generating monetary income, getting rid of fossil fuels will have a lesser impact now than it had two decades ago.

Unfortunately, nothing could be farther from the truth. Such arguments are akin to believing that if a modern plane can fly 5,000 miles with four jet engines turned on, and 6,000 miles with just two engines, it will fly 7,000

---

[8] From Greek: *petra* (rock) + Latin: *oleum* (oil).

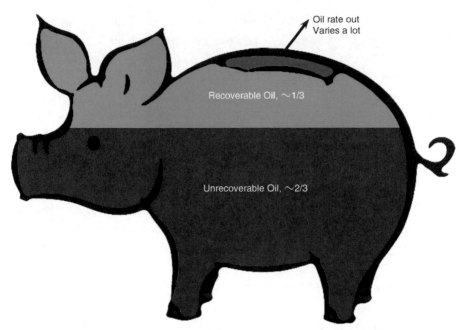

Accumulation = Piggy Bank, Coin Slot = Oil Wells, Injection Wells, and Surface Facilities

Fig. 3.7 Petroleum – or crude oil – fills the voids in layers of porous permeable rock, such as sand, sandstone, or limestone. These rock layers, "oil reservoirs," are usually confined from above by impermeable shale or clay layers. Only about a third of the petroleum in the rock pores can be recovered, unless special and expensive measures are undertaken. These special measures, also called "enhanced oil recovery (EOR) methods," include injection of water, surfactants, carbon dioxide, or heat to mobilize and push out the oil. After a successful EOR process, 50–70% of the oil in place is recovered

miles with all engines turned off. In the late nineteenth century, this type of thinking was known as the "Jevons paradox," after the British economist William Stanley Jevons, who posited that increased efficiency causes more energy use, not less.[9] This paradox is discussed in greater detail later in the book. In essence, modern economies are ever more dependent on energy, and

_____

[9] Suppose that you drive a Toyota Prius that allows you to save half of the money you were spending earlier on transportation. Unless you use the saved greenbacks to ignite wood in an efficient stove, or give them to the poor in Haiti, you will end up buying extra goods and services from less efficient sources, and you will cause more fossil energy to be burned. This way, your cumulative use of fossil fuels will actually increase, even as your rate of using these fuels directly decreases. This is the fundamental quandary facing the "decarbonizing" California.

the same quantity of energy supports more activities today than it did yesterday. Eliminate petroleum, coal, and natural gas, and nothing will be left of our current way of life, and two third or 67% of the current human population will have to perish. The latter estimate can be arrived at by extrapolating human population growth between 1650 and 1920–2010.[10] One can conclude that only 2.2 billion people could subsist mostly on plant carbon, but use some coal, oil, and natural gas. Therefore, it is reasonable to say that today 4.5 billion people (everyone born after 1940) owe their existence to the industrial Haber–Bosch ammonia process that produces nitrogen fertilizers from coal and natural gas and the fossil fuel-driven, fundamentally unsustainable "green revolution," as well as to vaccines and antibiotics.

Since the industrial revolution, a dramatic expansion in the scale of human endeavors, and of the human population, has expanded the human economy out of proportion to the economy of the Earth. This growth was mostly subsidized by the availability of cheap abundant carbon that underpinned all technological advances. It is simply too late for humankind as it exists today to go back to a state of harmony with the ecosystems we inhabit. Individual societies and cultures embark on paths from which there is often no return. A long time ago this meant distinct evolutionary paths for Roman society and Byzantine society. These paths and their environmental impacts were different from those of the Slavs, Vandals, Mayas, or Aztecs. Modern industrial society is not alone in grappling with this energy–complexity spiral, as we explore in Chap. 6.

Today, for the first time in human history, almost all of humankind follows in lockstep along a single path of industrialized globalized economy and culture. Regardless of where this path may lead, it is simply impossible to leave it today and keep current global capitalism intact and thriving. In other words, in 2011, there is no institutional will to make changes to the present course, and there is little social appetite to change our mostly comfortable lives.[11] Change will be forced upon us only by a decreasing rate of supply of cheap carbon. This point is key when we talk about offshore drilling.

---

[10]The early agrarian revolution in Europe, as well as the opening of the Americas was well under way by 1650. Clean drinking water supplies spread by the mid-nineteenth century.

[11]A quick Google search on January 14, 2011, yielded the following results: "global warming," 23 million hits; "climate change," 34 million hits; "Paris Hilton," 37 million hits; and "iPod," 262 million hits.

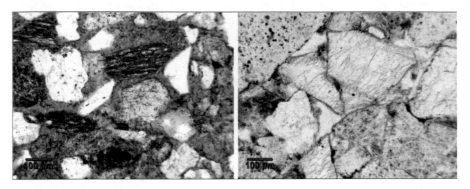

Fig. 3.8 Two thin sections showing under a microscope a high-quality reservoir sand-stone on the left and a nonreservoir sandstone on the right. The scale is 100 microns, the diameter of a typical medium sand grain. Both images are from the Alaska Peninsula–Bristol Bay region. The high-quality reservoir sandstone from the Bear Lake formation has abundant interconnected pores filled with the blue resin. These pores are capable of storing oil and gas and allowing these fluids to flow to producing wells. The nonreservoir sandstone from the Naknek Formation consists of the tightly fitted sand grains and cement-filled pores. It has essentially no porosity and no permeability. (Source: energy-alaska.wdfiles.com/local--files/natural-gas-as-a-resource/natural_gas171.jpg)

It Is the Rate, Stupid!

Hydrocarbon resources in oil and gas fields are geological accumulations of chemically transformed marine plants that can be more than two billion years old, although most petroleum is "younger" than 20 million years. Oil and gas do not reside in underground caves, lakes, or rivers, but in the microscopic pores of sedimentary rock, as illustrated in Fig. 3.8. This rock was deposited in massive layers, and was often chemically altered at depth under high pressure and temperature. Therefore, the oil companies do not tap into underground rivers of oil by opening a big tunnel to the surface, but have to extract oil and gas that move extremely slowly, if at all, through the tiny tortuous openings in the rock that are connected over hundreds of meters. Oil production is increasingly difficult and requires great skill, huge capital outlays, and cutting-edge technology. The oil and gas industry is by far the largest of all human technical endeavors, and Exxon Mobil, Royal Dutch Shell, BP, Saudi Aramco, and the like, are among the largest corporations on Earth. In 2009, Walmart, which is comparable in financial size to Exxon Mobil, employed 2.2 million people worldwide, whereas Exxon has a mere 91,000.

Petroleum has accumulated over millions of years from the almost imperceptible annual deposits of marine biomass, which in turn, like ourselves, is converted solar energy. You can think of petroleum resources as a huge global banking account. Up until recently our withdrawal centers (ATMs), that is, oil and gas wells, operated with few if any restrictions. A good well in Saudi Arabia might produce 10,000 barrels of oil per day. Similarly, a good well in the Gulf of Mexico can deliver 5–10 thousand barrels of oil per day. But, with time, Mother Nature imposes the ever more stringent daily withdrawal limits on our ATM cards. Old and easy oil reservoirs are depleted, the new ones are less productive, and we need more wells to produce less oil per day at a higher monetary and energy cost.

There are plenty of fossil fuels ("resources") left everywhere on the Earth. The resource size (current balance of a global banking account) is confused with the speed of drawing it down (allowed daily ATM withdrawals). It is the total rate of resource production that peaks, not the resource size which is gigantic but mostly impossible to recover at economically feasible rates.

The International Energy Agency, IEA, is an intergovernmental organization that acts as energy policy advisor to 28 member countries.[12] As with every political body, IEA has been in denial about the peak of oil (and coal and gas) production for many years. To IEA's credit, in 2008, they showed the global oil production rate estimate reproduced in Fig. 3.9. This estimate clearly shows a peak of oil production from the existing oil fields starting in 2004, just as many of us predicted years ago.[13] To be consistent[14] with their previous oil consumption estimates, IEA has created a magical wedge of oil that is yet to be found or produced. According to IEA, this wedge would

---

[12] See www.iea.org/country/index.asp for the list of the countries, mostly European and all developed.

[13] "Ring the bells that still can ring/Forget your perfect offering/There is a crack in everything/That's how the light gets in," sang Leonard Cohen encouragingly in his *Anthem*.

[14] In most corporate organizations, consistency and timeliness matter, but accuracy does not. Steady demand for petroleum is much easier to extrapolate into the future than supply. One starts with the current consumption rate of petroleum and ratios it simply with the growing population, while making special allowances for the fastest growing economies, such as China, Brazil, or India. In contrast, prediction of oil supply is one of the most complex activities of a modern society and has inherent large inaccuracies. Economists routinely equate energy demand with energy supply. This is wrong and when imaginary bets meet real resources the outcome can be disappointing.

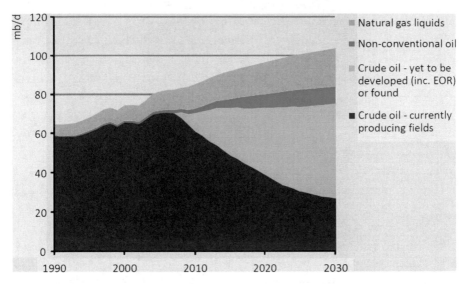

Fig. 3.9 According to the International Energy Agency (IEA), there is a peak of produced oil and 50 million barrels of conventional oil per day (mb/d) will be missing by 2030. IEA promises to fill this gap with new discoveries, further development of existing fields, enhanced oil recovery (EOR), such as CO_2 or steam injection, and a further increase of production from the Canadian tar sands. All these alternatives to conventional oil will alleviate the shortages somewhat, but their aggregate cannot possibly make up for the declining oil production in the existing fields. Note that natural gas liquids (NGL) cannot power a majority of the cars currently on the road, and will not play a major role globally as a source of liquid transportation fuels, unless there is a massive conversion of gasoline-burning engines to NGL-burning ones. As shown in Fig. 3.10, all EOR projects in the world currently deliver only 2.5 million barrels of oil per day, and the Canadian tar sands deliver 1.5 million barrels per day of synthetic crude oil and bitumen. (Source: www.iea.org/speech/2008/Tanaka/cop_weosideeven.pdf)

deliver about 50 million barrels of oil per day 20 years from now. In 2009, enhanced oil recovery (EOR) was about 2.5 million barrels of oil per day from hundreds of projects worldwide, half of which are found in the United States. It is impossible for EOR to deliver an oil production rate five to ten times higher than the current rate. A separate wedge for tar sands might provide another 7–10 million barrels per year for the next 20 years, says IEA. Such a high production rate would be several times more than the current 1.5 million barrels of synthetic crude oil and bitumen produced in Athabasca, Canada. As illustrated in Fig. 3.10, this too is impossible.

The only truly growing oil and gas production has been coming from offshore projects. Some of these projects, like those in the deep Gulf of Mexico, can deliver very high rates of oil production and quench our insatiable thirst for petroleum for another decade or so. This fact is at the core of

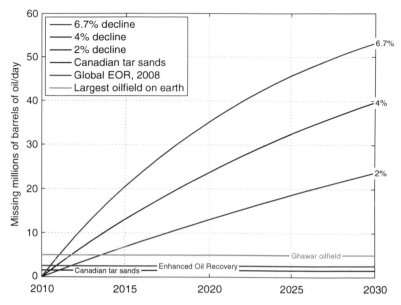

Fig. 3.10 In their 2008 World Energy Outlook, the International Energy Agency stated that we might have a 6.7% decline of the global oil production rate, not the mere 2% or 4% also shown here. Ghawar in Saudi Arabia (the top horizontal line) is the largest oil-producing oilfield in the world. The current global enhanced oil recovery and Canadian tar sand production are shown as the middle and lowest horizontal lines, respectively. We will have to add an equivalent of the Ghawar every 1–4 years, depending on the rate of oil production decline, and this is clearly impossible!

permitting the deep offshore projects as quickly as possible, or even quicker. The U.S. government and its constituents need both the oil production rate and the revenue from the deepwater offshore projects.

Inasmuch as the real production rate of liquid petroleum is peaking and the imaginary additions are unlikely to make up for the rate deficit, the world, but especially the developed countries, and above all the United States, will face inevitable shortages of liquid transportation fuels. For the completely unprepared United States, such shortages may be economically devastating.

So why did the global rate of oil production peak?[15] Ever since M. King Hubbert's seminal work in the mid-1940s, there has been an ongoing controversy about the existence of "Hubbert cycles," and their ability to predict the future rate of mining a natural resource. The Hubbert cycle is a bell-shaped curve of resource production, which postulates that production

---

[15] According to some, the global production rate of conventional petroleum has entered a 6-year-long "undulating plateau" of about 72 million barrels per day.

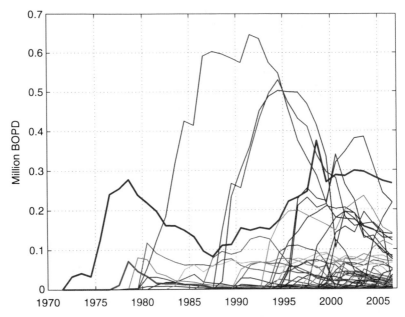

Fig. 3.11 Oil production rate in barrels of oil per day (BOPD) from each of the 65 North Sea and Norwegian Sea oilfields in Norway can be treated as an independent random variable. Their sum, the total production rate, is then a random-sum process and it should yield a bell curve (a Gaussian distribution) by the central limit theorem of statistics. The thicker lines refer to the oil production rates from Ekofisk and West Ekofisk, one of the largest oilfields in the North Sea. Note that these curves are neither symmetrical, nor do they have just one peak. (Data sources: the *Oil and Gas Journal*, OG&J, and the Norwegian Government (2009))

reaches a peak and then must decline. It has been demonstrated in several instances to be an accurate description and predictor of rising and falling production. This controversy seems to have been more ideological than scientific.[16] The inevitable emergence of Hubbert cycles which approximate the evolution of the total production rates from populations of coal mines, oil reservoirs, ore deposits, U.S. patents, soil productivity, and so on, was put to rest in the notes to a course offered at Berkeley by Patzek in 2007, and in a more recent paper by Patzek and Croft in the *Energy Journal* in 2010. This paper is listed in the Further Readings at the end of the chapter.

So far, the best visual example of emergence of a Hubbert curve consists of 65 offshore oil fields in Norway, whose production rates are shown in Fig. 3.11.

---

[16] For example, the long-standing religious conviction of the Cambridge Energy Research Associates (now IHS CERA) that a peak of oil production rate must not exist, has yet to be challenged by facts on the ground.

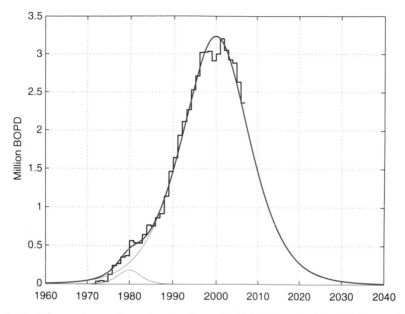

Fig. 3.12 The total rate of production from the fields shown in Fig. 3.11 is an almost perfect bell curve, called here a Hubbert curve or Hubbert cycle. The peak production rate is predicted for the year 2000, at 3.25 million barrels of oil per year. (Data sources: OG&J and the Norwegian Government (2009))

After a simple summation of the tangled spaghetti of individual field production histories, there emerges a picture-perfect bell curve or "normal distribution" in Fig. 3.12. The cumulative production is matched perfectly and the predicted ultimate oil recovery is 26 billion barrels, as shown in Fig. 3.13.

Norway has been a unique human experiment; it had no wars, revolutions, or great depressions over the time period of the North Sea development. What did vary, and wildly, was the price of oil. That variation apparently did not curtail oil production, nor did it distort the Hubbert cycle. Mixing Norwegian Sea and North Sea production also was not a problem; these are two different areas, but similar geology and operating conditions caused them to act in a single cycle. The most serious, but still slight, deviation between oil production and the Hubbert cycle was caused by very low oil prices between 1995 and 2000.

One should not read into the Hubbert cycles more than is justified. The production rate from each oil field, gas field, or coal mine approximates a random variable in time. A given population of these random variables sums

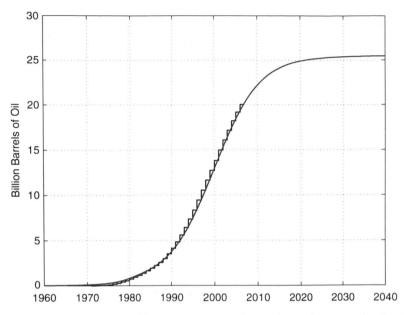

Fig. 3.13 The cumulative Hubbert curve versus the total cumulative production from the fields shown in Fig. 3.11. The cumulative oil production from the Norwegian oil-fields is predicted to be around 26 billion barrels. (Data sources: OG&J and the Norwegian Government (2009))

up to a bell curve. At later times, new fields are produced with technologies that were not previously available. The production rates from these new fields sum to the new bell curves. As long as the individual random variables emerge from optimized[17] exploitation of a resource and informed decisions are made to maximize ultimate recovery from each field project, the Hubbert curves are fairly representative of the best efforts of geology, science, engineering, and economics at hand.

Of course the existing Hubbert cycles cannot foresee the future ones. Thus the ultimate recovery of a resource will always be larger than the sum of the current, well-established cycles. Just how much larger? This we do not know, but the discrepancy between what has been and is extrapolated into the future, and what can be expected, ought to decrease with time because the Earth is finite.

---

[17] In this context, "optimized" means using the best design and technology available to the highest extent possible.

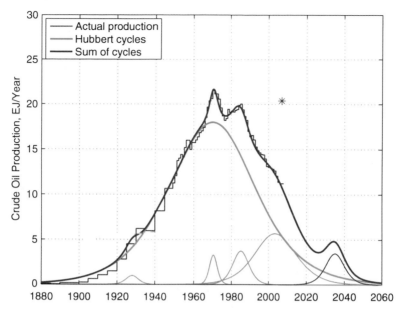

Fig. 3.14 The Hubbert cycle analysis of U.S. crude oil/condensate endowment. The main cycle gives the original Hubbert estimate of ultimate oil recovery of 200 billion bbl. The smaller cycles describe the new populations of oil reservoirs (Alaska, Gulf of Mexico, Austin Chalk, California Diatomites, etc.) and new recovery processes (waterflood, enhanced oil recovery, horizontal wells). Note that the total rate of production of all oil resources in the United States goes through a peak, and cannot continue growing exponentially. In fact, in 2003, total U.S. oil production decreased all the way down to the 1950 level. The star shows the higher heating value of automotive gasoline burned in the United States in 2007. The Hubbert cycles shown here were fixed in 2001, and continued to predict the U.S. oil production for another nine years. Also shown, as the rightmost small Hubbert cycle, is a hypothetical production of the undiscovered, technically producible 7.7 billion barrels of oil from Area 1002 of the Arctic National Wildlife Refuge, Alaska, (ANWR) that peaks in 2035. (Data sources: US DOE EIA, USGS, State of Alaska)

Technology

We have a quasi-religious faith that technology will always save us from difficulties. In fact, Figs. 3.14 and 3.15 show us that better technology has resulted in significantly improved petroleum recovery in the United States over and above the fundamental Hubbert cycle. In Fig. 3.15, the difference between actual cumulative oil production in the United States and the cumulative oil produced from the fundamental Hubbert cycle is at least 200 EJ,

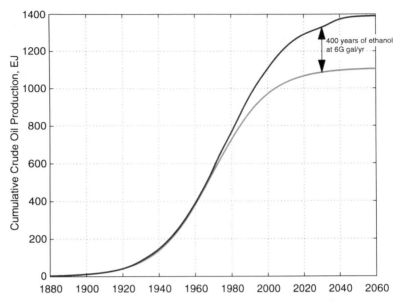

Fig. 3.15 The cumulative oil production in the United States cannot continue growing forever. The increment between the main Hubbert cycle and total recovery is equal to twice the total U.S. energy consumption in 2003. This difference is a contribution of new technology and research in the oil industry. Coincidentally, this difference is equal to 400 years of pure ethanol production at six billion gallons per year, which is the global record set in the United States in 2007. The small increase of the total recovery curve after 2030 is caused by a hypothetical production of 7.7 billion barrels of oil from the Arctic National Wildlife Refuge (ANWR) in Alaska

or 2 years of U.S. primary energy consumption.[18] This incremental oil has been – and will continue to be – produced mostly because of progress in well drilling and completion technologies (better and faster drilling methods, directional wells, horizontal wells, multilateral wells, smart multi-interval wells, etc.), fracturing technology, waterflood and polymer-enhanced water-flood, and enhanced oil recovery methods, mostly steam injection, CO_2 injection, and surfactant/polymer flooding. As we have seen in Chap. 2, offshore oil production contributes an increasingly larger share of the U.S. oil supply. At only 50 USD per barrel of oil, the value of incremental oil production in the United States alone would amount to 1.5 trillion in 2008 US dollars.

In 1954, the influential philosopher, Martin Heidegger, foresaw the logical progression of modern civilization by describing technology as a

---

[18]The ultimate difference will be at least 400 EJ.

"standing-reserve" of energy for humans to order nature and, in turn, be enframed and perhaps consumed by their technology.

Heidegger can be translated into plain English as follows.

- We are an impatient species that regards a standing-reserve of energy as a must.
- Because we cannot control technology, technology cannot be our tool to control nature.
- We tend to think of technology as an instrument that is outside us. Instead, we are part of a bigger system that comprises us and technology.

Our technological civilization is all about power (the rate of energy use). This power is extracted from the earth with the ever-improving technology applied to unlock more fossil fuels. With time, our technology might condition our thinking so completely, that we would not be able to see the earth as more than a source of energy. If you doubt this, a proposed mine in West Virginia, which would have buried miles of Appalachian streams under millions of tons of rock, and obliterated a healthy environment and drinking water supply for thousands of people, has been the subject of ongoing controversy and litigation. The mining company and politicians in West Virginia expressed fury at an attempt by the EPA to stop the mine construction, focusing on the economy, but not seeing the potential for an environmental catastrophe.[19]

Here is how Kurt Vonnegut described in 1973 what happened to West Virginia ("Breakfast of Champions," Chap. 14, p. 123):

> The surface of the State had been demolished by men and machinery and explosives in order to yield up its coal. The coal is mostly gone now. It had been turned into heat.
>
> The surface of West Virginia, with its coal and trees and topsoil gone, was rearranging what was left of itself in conformity with the laws of gravity. It was collapsing into all the holes which had been dug into it. Its mountains, which once found it easy to stand by themselves, were sliding into valleys now.
>
> The demolition of West Virginia had taken place with the approval of the executive, legislative and judicial branches of the State Government, which drew their power from the people.
>
> Here and there an inhabited dwelling still stood.

---

[19] See. e360.yale.edu/content/feature.msp?id = 2198

Energy and Innovation

We are often assured that innovation in the future will reduce our society's dependence on energy and other resources while providing a lifestyle such as we now enjoy. We discuss this point further in Chaps. 5 and 9. Here we observe that rates of innovation appear to change in a manner similar to the Hubbert cycle of resource production. This finding has important implications for the future productivity and complexity of our society.

Energy flows, technology development, population growth, and individual creativity can be combined into an overall "Innovation Index" which is the number of patents granted each year by the U.S. Patent Office per one million inhabitants of the United States of America. This specific patent rate has the units of the number of patents per year per one million people. Figure 3.16 is a decomposition of this patent rate into multiple Hubbert-like cycles between 1790 and 2009.

Interestingly, the fundamental Hubbert cycle of the U.S. patent rate peaked in 1914, the year in which World War I broke out. The second major rate peak was in 1971, coinciding with the peak of U.S. oil production. The last and tallest peak of productivity occurred in 2004. Note that without a new cycle of inventions in something, the current cycles will expire by 2050. In other words, the productivity of U.S. innovation will decline dramatically in the next 20–30 years, with some of this decline possibly being forced by a steady decline of support for fundamental research and development.

Energy and Complexity

Each new complex addition to the already overwhelmingly complex social and scientific structures in the United States is less and less relevant, while costing additional resources and aggravation. Most of this complexity is apparent to the naked eye: look at the global banking and trading system, the healthcare system, the computer operating systems and software, military operations, or government structures. The scope of the problem is also obvious in the production pains of Boeing's 787 Dreamliner, and in the drilling of the BP Macondo well.

We use more energy in more complex applications, and then need more complexity to manage the increased energy flows through society. In Chap. 4

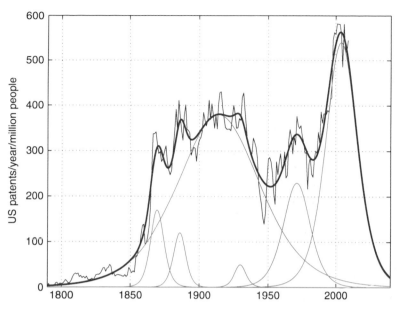

Fig. 3.16 The fundamental cycle of patented inventions in the United States peaked in 1914, when expressed per one million of people living in the United States. This was the classical science and engineering patent cycle. The first small peak in 1870, was related to the Civil War and newly acquired technological sophistication in the United States. The second small peak in 1885 was probably related to the innovators who were born post-1860 coming of age. The third small peak in 1930 was a boost to innovation during the roaring 1920s. Curiously, the third largest patent cycle, mostly in engineering, peaked in 1971, the year in which U.S. oil production peaked too. The latest and largest patent cycle, with an apparent peak in 2004, describes mostly progress in biotechnology, medicine, materials science, and informatics. Note that this latest patent cycle is projected to decline at a 13%/year rate. This cycle has little to do with classical geosciences and engineering applied to hydrocarbon systems. In geological sciences and engineering, we are already at a point of significantly diminishing returns, and breakthroughs will not be realized by continuing business as usual. Radically new solutions must be found in our quest for cheap abundant energy. We are not talking here about solar photovoltaics and wind as these technologies have rapidly matured as well. (Sources: U.S. Patent Office; U.S. Census Bureau; and Strumsky et al., Complexity and the productivity of innovation, *Systems Research and Behavioral Sci. Syst. Res.* 27, pp. 496–509, 2010)

we see how this complexity has been manifested in offshore drilling structures and operations. In Chaps. 5, 6, and 9 we discuss complexity as a society-wide phenomenon, how it comes about, what it costs, and what it means for our future.

Further Reading

1. Baum, A.W., Patzek, T.W., Bender, M., Renich, S., Jackson, W.: The visible, sustainable farm: a comprehensive energy analysis of a midwestern farm. Cr. Rev. Plant Sci. **28**(4), 218–239 (2009). gaia.pge.utexas.edu/papers/TheSunshineFarm.pdf. Presents the most detailed to-date discussion of energy and mass flows through an organic farm and shows how competitive this farm is relative to industrial farms
2. Brown, J.H., et al.: Energetic limits to economic growth. BioScience **61**(1), 19–26 (2011). ISSN 0006–3568, electronic ISSN 1525–3244, 2011, news.unm.edu/wp-content/uploads/2011/01/Brown-BioScience-2011.pdf. Shows that the monetary "metabolism" of an economy scales with energy throughput just like animal metabolism with body weight
3. Holland, J.H.: Emergence: From Chaos to Order. Addison-Wesley, Reading (1998)
4. Allocation of the Holy Father John Paul II: In the *Emergence of Complexity in Mathematics, Physics, Chemistry, and Biology*, Page 465–468, Pontifical Academy of Sciences, (1992)
5. King, F.H.: *Farmers of Forty Centuries or permanent agriculture in China, Korea and Japan*, Published by Mrs. King, F.H. 1911, also published on the web, courtesy of Steve Solomon, www.soilandhealth.org. The best description of a national organic agriculture 50 years before the term was even known
6. Lambert, F.: *Shakespeare and Thermodynamics: Dam the Second Law!*, 2008, shakespeare2nd-law.oxy.edu. Probably the best, easy-to-understand discussion of the Second Law of Thermodynamics
7. Morowitz, H.J.: The Emergence of Everything – How the World Became Complex. Oxford University Press, New York (2002)
8. Patzek, T.W., Croft, G.: A global coal production forecast with multi-Hubbert cycle analysis. Energy **35**, 3109–3122 (2010). gaia.pge.utexas.edu/papers/EnergyCoalPaperPublished.pdf. Nice examples of the emergence of Hubbert peaks and their implications for the global rate of coal production and carbon dioxide emissions
9. Patzek, T.W., Croft, G.: Supporting online materials to a global coal production forecast with multi-Hubbert cycle analysis. Energy **35**, 35 (2010). gaia.pge.utexas.edu/papers/EnergyCoalPaperSOM.pdf
10. Patzek, T.W.: Thermodynamics of agricultural sustainability: the case of US maize agriculture. Critical Reviews in Plant Sciences **27**(4), 272–293 (2008). gaia.pge.utexas.edu/papers/Thermodynamics of Agricultural.pdf. An in-depth discussion of unsustainability of modern agriculture and available remedies
11. Patzek, T.W.: Exponential growth, energetic Hubbert cycles, and the advancement of technology. Arch. Min. Sci. **53**(2), 131–159 (2008). gaia.pge.utexas.edu/papers/PatzekManuscriptRevised.pdf. A nice discussion of unconventional hydrocarbon resources and their impact on the U.S. economy
12. Patzek, T.W.: *How Can We Outlive Our Way of Life?* Paper prepared for the ministerial OECD round table on sustainable development, "Biofuels: Is the cure worse than the disease?" OECD Headquarters, Château de la Muette, 2 rue Andre Pascal, 75016 Paris, 12 Sept 2007 www.oecd.org/dataoecd/2/61/40225820.pdf. Patzek's second most popular paper

13. Patzek, T.W.: OECD dinner speech at the ministerial OECD round table on sustainable development, 11 Sept 2007 gaia.pge.utexas.edu/papers/OECD09112007TalkFinal.pdf. Everyone stopped eating during that speech and you could hear a mosquito flying
14. Patzek, T.W.: *The Real Biofuel Cycles*, Online supporting material to science 312, 1747, 26 June 2006 gaia.pge.utexas.edu/papers/RealFuelCycles-Web.pdf
15. Patzek, T.W.: *Thermodynamics of the Corn Ethanol Biofuel Cycle*, Invited paper in the special issue of Critical Reviews in Plant Sciences **23**(6), 519–567 (2004) gaia.pge.utexas.edu/papers/CRPS416-Patzek-Web.pdf. Patzek's most popular paper ever
16. Snow, C.P.: The Two Cultures and the Scientific Revolution. Cambridge University Press, London (1959)
17. Strumsky, D., Lobo, J., Tainter, J.A.: Complexity and the productivity of innovation. Systems Research and Behavioral Science **27**, 496–509 (2010). The most in-depth analysis of the inter-dependence of patentable innovation and complexity
18. Yale Environment 360, *Leveling Appalachia: The Legacy of Mountaintop Removal Mining*, focuses on the environmental and social impacts of mountaintop removal and examines the long-term effects on the region's forests and waterways. e360.yale.edu/content/feature.msp?id = 2198

Chapter 4

Offshore Drilling and Production:
A Short History

Drilling in Louisiana's marshes and shallow waters is as old as anyone there can remember, and – for better or worse – the expanding presence of the oil and gas industry has changed everyone's lives. An oral history[1] captures the richness and complexity of interactions between people and the technology that invaded the small fishing and shrimping communities. Here is but one short excerpt.

> Okay, my name is Pershing J. Lefort. I was born in 1924 and so I was 5 years old when the stock market dropped overnight and then the Depression years beginning from that point. In 1934, I was 10 years old. And oil researchers had discovered shallow oil deposits in the Golden Meadow area and therefore many small companies brought land drilling equipment and started to develop the field in the early thirties. By the time I was 10 years old they had a few wells drilled and I took a lot of interest in the type of work these men were doing and of course these men were mostly men that came from west Texas with some knowledge of drilling and road building and stuff of that nature. I was completely surrounded by little drilling rigs. They was small rigs and the wells they were drilling were shallow. All free flowing wells. And these men being working people away from home having kids probably my age, some. They were good boys. They'd bring us on the rigs and then some time my mother would make a pot of coffee and I'd bring it down and put a pot of coffee on the rig. I'd even sit with the driller and put my hand on the brakes just for the fun. In those days there was no safety practice done and you

[1] McGuire, T. 2004. History of the offshore oil and gas industry in southern Louisiana: Interim Report; Volume II: *Bayou Lafourche – An Oral History of the Development of the Oil and Gas Industry*. U.S. Dept. of the Interior, Minerals Management Service, Gulf of Mexico OCS Region, New Orleans, LA. OCS Study MMS 2004-050. 148 pp.

know anything went. So I got to feel the rig and watch those drillers when I was quite young. My parents leased their property to a company named Brown and Root and they drilled an oil well right off the house in the back and another one later right in front of the house which was the biggest mess. I remember it being a big mess but I had fun being on these rigs with these men. It was the biggest mess because in those days there was no regulation for drilling mud control or venting on gas, you just brought the oil in, put it in the small storage tank, 500 or less barrels and no pipeline. The well would only be open a few hours before he'd fill the tank then had to wait for a barge. The only way of transporting oil out of east Golden Meadow and west Golden Meadow as well was by barging, using Bayou Lafourche. That's what we had to use Bayou Lafourche to cross to go to school. We didn't have no school, no stores; we didn't have nothing. We lived in the marsh. So it was quite an exciting life when I look back. I had a real ball. Then it got so bad; Texaco had a well that blew out on the west side of Golden Meadow. The [large salt dome] that our field is on has salt and sulfur. Therefore that gas is very toxic but it didn't matter. They just vented it through pipes and ventilated and it burned. There was no way to handle gas and the companies had no interest in gas. It was the heavy crude they wanted to be able to sell to refineries.

When Mr. Lefort was telling his story, the U.S. oil and gas industry was already in its adolescence. For the beginning of modern petroleum production in the U.S., we have to go back another half century to the small town of Titusville, Pennsylvania.

Although crude oil was scooped from natural seeps or shallow pits dug into the ground for thousands of years, it was not until August 27, 1859, when Colonel Edwin Drake of Seneca Oil drilled into an oil field near Titusville, Pennsylvania. As depicted in Fig. 4.1, Drake used sections of iron pipe with a total length of 10 meters (32 feet), driving them all the way down to bedrock. The iron pipe prevented the borehole from collapsing and Drake could safely drill inside it. He derived no profit from his epochal achievement and died penniless two decades later.

Vanity Fair (Fig. 4.2) later commemorated Drake's breakthrough in a cartoon that depicted sperm whales celebrating because their oil was no longer needed to light America's lamps. Today, after many improvements, Drake's basic technique is still familiar to the millions who listened to news reports as efforts were made to stem the flow of oil into the Gulf of Mexico.

Exploration and drilling for oil over open water began 40 years later, when wells were drilled from piers extending off the beach at Summerland, California. Gulf Oil drilled the world's first truly "offshore" oil well, fully detached from land, in inland waters at Caddo Lake in 1911.

In the 1920s, when drilling began in the thick marshes and wooded swamps of Louisiana's bayou country, crews turned to methods and equipment

Fig. 4.1 By August 27, 1859, in the sleepy lumber town of Titusville, Pa., "Colonel" Drake hammered 10-foot sections of an iron pipe into the ground and, at a depth of 21 meters below ground, he finally struck oil. The world changed forever. Over a century and a half, Drake's 25 barrels-per-day well would give rise to the world's largest industry. Globally, the industry now produces 73 million barrels per day, making oil the world's most strategic commodity, supplying 40% of the world's energy. (Source: evworld.com/article.cfm?storyid = 1749)

used by muskrat trappers. The trappers relied on flat-bottomed *pirogues* to navigate *tranasses*, trapping ditches hacked out to provide marsh access. Only wide enough to accommodate a pirogue, these passageways were used to explore the marsh and represented the first attempt to dissect it with a network of canals. New channels were added daily, but old ones only rarely filled with sediment. Once a trananasse was cut, it remained for years, often enlarging into a bayou. Many channels that began as pirogue trails evolved through repeated use, storms, and current flow into permanent features. These watercourses have affected drainage patterns, influenced water salinity, and contributed to marsh deterioration.

Exploring for oil in marine marshes tended to be a slow process, involving the adaptations of land-based equipment and technologies to particular

GRAND BALL GIVEN BY THE WHALES IN HONOR OF THE DISCOVERY OF THE OIL WELLS IN PENNSYLVANIA.

Fig. 4.2 The introduction of iron drill pipes was key to striking oil near Titusville, Pennsylvania, on August 27, 1859. Jubilant sperm whales, depicted by Vanity Fair on April 20, 1861, thanked Colonel Drake and Seneca Oil for saving them from extinction. Almost overnight, kerosene (paraffin oil) became a cheap abundant alternative to blubber oil. The truth, however, was more complicated. In spite of the impossibility of replacing geological carbon accumulations (fossil fuels) with annual carbon flows though flora and fauna, the indiscriminate exploitation of ecosystems will continue because large monetary investments are never abandoned of free will. With prior investments in steel-hull, steam-powered, cannon-equipped ships, for example, whale hunting continued until sperm whales neared extinction. Fast forward 150 years. By 2008, gigantic agrofuel plantations have invaded land in the tropics, destroying some of the most important ecosystems on the planet

locations. Although drilling in open waters of the Gulf of Mexico had been taking place for some time, a historical milestone for the offshore industry was reached in 1947, when Kerr McGee completed the first offshore well out of sight of land. Today's deepwater wells like the ill-fated Macondo are its direct descendants, and when offshore drilling moved into deeper waters of up to 100 feet, fixed platform rigs were built.

Figure 4.3 shows a few examples of the evolution of offshore oil and gas drilling and producing structures. Several other types of drilling and production platforms used in the Gulf of Mexico, and four examples of

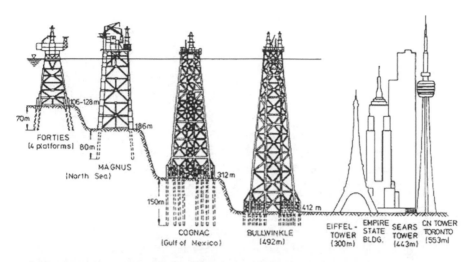

Fig. 4.3 Types of offshore oil and gas structures include (*Bottom figure from left to right*): Two conventional fixed platforms (deepest: Shell's Bullwinkle in 1991 at 412 meter/1,353 feet, GOM); a compliant tower (deepest: ChevronTexaco's Petronius in 1998 at 534 meter/ 1,754 feet, GOM); two vertically moored tension leg and mini-tension leg platforms (deepest: ConocoPhillips' Magnolia in 2004, 1,425 meter/4,674 feet, GOM); a spar platform (deepest: Dominion's Devils Tower in 2004, 1,710 meter/5,610 feet, GOM); two semi-submersibles (deepest: Shell's Perdido in 2010, 2,438 meter/8,000 feet, GOM); a floating production, storage, and offloading facility (deepest: 2005, 1,345 meter/4,429 feet, Brazil); a sub-sea completion and tie-back to host facility (deepest: Shell's Coulomb tie to NaKika 2004, 2,307 meter/ 7,570 feet. In 2009, the deepest reeled flowline installation at a water depth of 2,961 meter (9,713 feet) was Shell's Perdido, GOM). Sources: Office of Ocean Exploration and Research, National Oceanic and Atmospheric Administration; Shell

complete offshore installations, are described for the interested reader in Appendix B. Appendix C summarizes the extensive personnel requirements for a complex offshore platform.

Fig. 4.4 A string of marine riser pipe. Note the small pipe openings embedded in the riser collar. These small pipes are used to control the well ("choke" and "kill" lines), and deliver hydraulic fluid to the BOP rams and LMRP annulars. In addition, the riser pipes carry power and electronic communication with the well. (Image source: Wikipedia)

How to Drill a Well in Deep Water

Offshore wells are drilled in much the same way as their onshore counterparts. A conduit made from lengths of steel pipe permits drilling fluids to move between the drillship at the water surface and the seafloor. This conduit is called a "riser" (Fig. 4.4). For all of you nontycoons out there, a glossary of technical drilling and production terms can be found in Appendix A.

In deep water, the drillship is kept in place by a dynamic control system that uses a Global Positioning System (GPS), water jets, and propellers. The riser is fitted with ball-and-slip joints that permit a one-mile-long string of riser pipe to move up and down and bend slightly with the wave-induced movement of the ship.

An offshore well is drilled using a set of slender steel pipes and other tools that, connected through joints, comprise a "drill string." A drill bit assembly is attached to the bottom of the string of pipes. Heavy sections of pipe, called "drill collars," add weight and stability to the drill bit. Each ordinary pipe in the string is about 30 feet/9.1 meters long and weighs about 600 pounds/270 kilogram; drill collars can weigh in excess of 4,000 pounds/1,800 kilogram per 30-foot string. As drilling proceeds, and the well gets deeper, the drilling crew adds new sections of drill pipe to the ever-lengthening drill string. Hydraulic devices keep constant tension on the drill string to prevent the motion of the ship and riser from being transmitted to the drill bit.

The drill string is lowered through the riser to the seafloor, passing through a system of safety valves called a "blowout preventer" (BOP), and the "lower marine riser package" (LMRP), shown in Fig. 4.5. This stack of

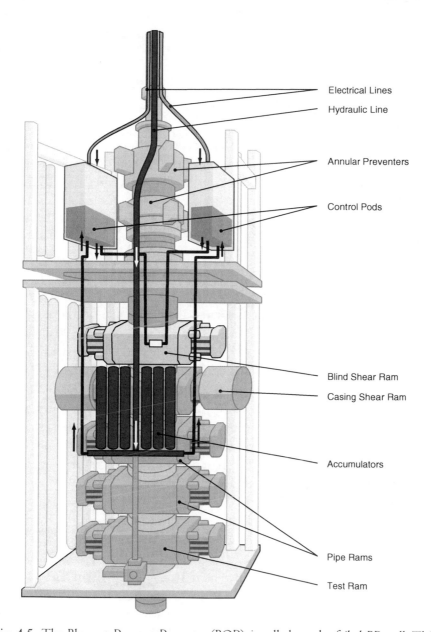

Electrical Lines

Hydraulic Line

Annular Preventers

Control Pods

Blind Shear Ram

Casing Shear Ram

Accumulators

Pipe Rams

Test Ram

Fig. 4.5 The Blowout Pressure Preventer (BOP) installed on the failed BP well. This 450 t plus stack of mechanical devices, powered by electricity and hydraulic systems, is attached to the wellhead at the bottom, and to the umbilical 18–3/4 inch. ID riser pipe at the top. The riser pipe is allowed to sway by the Lower Marine Riser Package (LMRP) at the top of the BOP and a ball-and-slip joint at the drillship. In the outer shell of the connected marine riser pipes, there are embedded hydraulic lines, communication lines, and electrical cables that connect the ship operator with the BOP. In case of emergency: (1) the operator sends an electrical signal to one of the two pods that control the BOP; (2) the pod activates a hydraulic system connected to the ship and/or local canisters; (3) a "shuttle valve" directs the high-pressure hydraulic fluid to the blind shear ram; and (4) the ram closes and cuts through a pipe inside it; the well is shut in, or so we think

multiple safety valves ("rams" and "annulars") is designed to contain any back flow of high-pressure fluids (especially gas) the drillers might encounter in an open borehole that penetrates the overpressured rock strata. When it works, the BOP prevents a possible well "blowout," an uncontrolled eruption of the high-pressure oil, gas, or wellbore fluids. The LMRP serves to connect and disconnect the riser pipe from the wellhead, and its annulars can stop the flow of fluids through the annular space of the riser.

The first stage of drilling is called "spudding" and starts when the drill bit is lowered into the seabed. We see in Fig. 4.6 that the bit can be of two types, a roller cone or rock bit which typically has three cones armed with steel or tungsten carbide teeth or buttons, or a diamond bit, impregnated with small synthetic "polycrystalline" diamonds. The drill bit is attached to a drill pipe (or a drill string) and rotated by a turntable on the platform's "derrick" floor. In addition, a "mud motor" attached to the drill bit can rotate it independently of the drill string rotation. Drilling with a mud motor is called "sliding," as opposed to the usual rotary drilling. Mud motors are used to drill slanted and horizontal wells. The drill bits range in diameter from 36 inches./91.4 centimeters (used at the start of the hole) to 8.5 inches./21.5 centimeters (used to penetrate the hydrocarbon reservoir).

Drilling fluid is pumped down the drill pipe and exits through nozzles in the drill bit at a high velocity. In offshore drilling, this fluid is usually an "oil-based mud" or a "synthetic mud." Oil-based mud uses diesel fuel as the base fluid. Oil-based muds are used for many reasons, such as increased lubricity, prevention of clay swelling, and greater cleaning abilities with less viscosity. Oil-based muds also withstand greater heat without breaking down. Synthetic-based mud uses synthetic oil as the base fluid. This mud is used most often on offshore rigs because it has the properties of an oil-based mud, but the toxicity of the fluid fumes is much less than that of an oil-based fluid. This is important when men work with the fluid in an enclosed tight space such as an offshore drilling rig. Oil-based and synthetic muds are densified by an addition of barite, or barium sulfate, a dense mineral. The mud density is usually measured in pounds per gallon (ppg). The density of sea water is 8.5–8.6 ppg, depending on water's salinity. Mud densities in excess of 16 ppg, or twice the water density, can be achieved.

A drilling mud serves several purposes. It raises the drill cuttings to the surface for disposal, provides the weight to keep the underground pressure in check, keeps the hole stable by caking the wall with a thin layer of clay, and cleans and cools the bit. The fluid is recycled through a circulation system

Fig. 4.6 From *left*: one tungsten carbide drill bit and two diamond-impregnated bits manufactured by Atlas Copco Craelius. (Source: Wikipedia)

mounted on the drilling rig that separates out the drill cuttings and allows the clean fluid to be pumped back down the hole.

Centralizers

The well "casing" is a slender steel tube used during the drilling of a well to prevent borehole collapse and protect the well's interior parts for several decades. Strong bow springs with hinged collars that go around the casing push it away from the borehole wall in all directions, thus centering the casing (Fig. 4.7). These spring assemblies are called "centralizers." When a casing is centered in a borehole, it is easier to place cement between the two, and seal off all paths for the flow along the casing. Otherwise, there is a high risk that a string of drilling fluid or contaminated cement is left where the casing contacts the borehole, creating a potentially disastrous channel for the flow of reservoir fluids upward.

Cement

The Presidential Commission Report of January 11, 2011 concluded that the most likely cause of failure of the Macondo well was leaky cement that never set properly. This cement was placed by Halliburton to seal the bottom

Fig. 4.7 *Left*: A typical casing centralizer with two collars that are screwed onto a string of casing tube. *Right*: A more sophisticated version of the centralizer on the *left* is a "centralizer sub," that can be screwed in-between two strings of casing, thus providing greater mechanical integrity (the centralizer collars cannot slip and lock casing in the hole at some intermediate depth). The BP engineers on the Deepwater Horizon did not like centralizers like the one on the left, thought that they did not have enough of the subs on the right, and installed only six centralizer subs, as opposed to 21 or so that were needed according to Halliburton. (Sources: Bond-Coat, Inc., Midland, TX; and Casing Systems International, LLC, Scott, LA)

189 feet of the 7-inch. production casing, and over 800 feet of the borehole outside the casing.[2] The special cement was puffed up with nitrogen bubbles to lower its density and avoid fracturing the rock around the well. This cement was supposed to seal the casing and the borehole permanently and, indeed, a proper cement seal would prevent both structures from being invaded by the high-pressure reservoir fluid.

---

[2]This design would place cement up to 500 feet above the uppermost hydrocarbon zone, violating BP's own rule of sealing at least 1,000 feet of annular space above the uppermost hydrocarbon zone.

Summary

Over the last 100 years, the world has seen a breathtaking expansion of complexity and size of offshore engineering and science. Tens of thousands of oil and gas wells have been drilled and completed in water depths ranging from a couple of meters to a couple of kilometers. In doing so, humanity has entered the realm of deep ocean, an environment much harsher and less forgiving than outer space.

We would like to think that taming the deep offshore environment is possible with sufficient technology In reality, it will always be a frontier struggle of humanity to grab treasures and run from the most inhospitable of environments. Although we hope that stricter regulations and control will avert future offshore disasters, no regulation will magically overrule the power of the ocean and the deep hot earth. With easy oil largely produced, supplies to meet demand must increasingly come from more remote, deeper, and often smaller reservoirs as we saw in Chap. 2. In the next two chapters, the resulting complexity and its implications for society take us far beyond oil drilling and production technology alone.

Chapter 5

The Energy–Complexity Spiral

Engineers build many wonderful things that few of us would choose to live without. Yet, as we have seen, some structures are of such complexity and magnitude that an unforeseen failure can kill nearly a dozen men, ruin thousands of livelihoods, and pollute a valuable ecosystem. Failure on this scale is obviously undesirable, yet it happens to bridges, space shuttles, and giant drilling rigs. In response, our instinct is to seek proximate causes, which include such factors as mistakes, oversights, and technical failures, the very things on which most attention has been concentrated in the news media. By applying some fixes – better training, better oversight, a different corporate culture – we assume that the accident could have been prevented and that we can avoid future ones. Engineers must examine and learn from these proximate causes of failure, but as a society we are bound to seek the ultimate cause of tragedies such as the Deepwater Horizon's blowout. The alternative is to lurch from failure to failure of increasing magnitude. We will find that the ultimate cause lies deep within humanity's history, and in the very essence of what it means to be a civilization. A civilization is a complex society, and complexity is a phenomenon that we must understand in order to comprehend our potential futures and shaping events such as the Gulf tragedy.

Human societies for at least the last 12,000 years after the invention of agriculture have grown progressively more complex in a manner that seems inexorable. This growth in complexity is closely linked to the growth in energy that is available to power our way of life. In this and later chapters, we examine how the very existence of the Deepwater Horizon is the inevitable result of this growth in complexity that has characterized so much of our history. So, too, is its failure.

J.A. Tainter and T.W. Patzek, *Drilling Down: The Gulf Oil Debacle and Our Energy Dilemma*, DOI 10.1007/978-1-4419-7677-2_5,
© Springer Science+Business Media, LLC 2012

In seeking the ultimate cause, our focus in this chapter is on complexity, its development in society, and the connection of complexity to energy. These are matters that most people do not think about very often, if at all. So we begin with a phrase that readers will find familiar: the complexity of modern life. A search for this term on Google.com conducted on July 10, 2010 returned 270,000 Internet pages. A lot of people apparently feel inspired, or perhaps compelled, to write about it. It is a term that does not need to be defined or explained. Nearly every member of industrial societies will understand it. We experience it in our daily lives. "The complexity of modern life" is not a complimentary or cheerful term. In fact it is pejorative. Even more, it is not just an impression or state of mind. Complexity has very specific consequences for society.

Although most of us cannot specify the details, we suspect that life was not always so complex. As the author known as PeakEngineer wrote in the *Energy Bulletin* (a website), "In the beginning, it was all so simple. Rub two sticks together, get a fire. Stick a pipe in the ground, get some oil. Trade a cow, get a llama." Life in the past may not have been quite this simple, but PeakEngineer's point is rhetorical and no one would contradict it. Life in the past was not simple, but in many ways it was certainly simpler. We pay a price for complexity, and two of the currencies for counting that price are stress and aggravation. Rubbing sticks together may be aggravating, but no doubt less so than the complexity of the many different activities that people juggle today.

The electronic devices on which we now depend have become some of the main culprits. There are always updates to download and install, viruses to avoid, or new software to master. Computers were supposed to save time and reduce effort. In November 2009, one of us (Tainter) bought a new cellular phone known as a "smart phone." It looked like it might make travel easier: get email away from home without having to carry along a laptop computer. Within hours the device started nagging to download updates and fixes. It continues to do so. If one does not comply with an imperative download, the device just continues to nag. Eventually it gets its way. The nagging is part of the cost of complexity.

The complexity of modern life naturally generates a backlash. The express form of that backlash is the so-called "slow movement." Like the complexity of modern life, it hardly needs to be explained. We understand it at first hearing. According to Wikipedia, "The **Slow Movement** is a cultural shift toward slowing down life's pace." It is intended to combat what is called "time poverty." The principle is simple: decelerate as much as possible. The dimensions of the movement include slow food, slow money, cittaslow (antihomogenization of towns and cities), slow parenting, slow travel, slow art, slow media, slow love, and perhaps more. There is even an International Institute of Not Doing Much.

The slow movement is actually one of several cultural movements that style themselves as the antithesis of modern life: the local currency movement, anticommercialism, antiglobalization, anti-ngenetically modified foods, communes, and so forth. Often adherents of these movements form religious sects, some temporary, others longer-lasting. The Amish come to mind.

These movements exist because complexity comes with a price tag. Although we do not usually think of that price in energy terms, in fact the complexity of modern life has an energy cost. A gallon of gasoline is said to be the energy equivalent of a person working 400 hours. The term "energy slave" is used to describe the energy equivalent of the physical work that a healthy youthful human could theoretically accomplish. Each European, by some calculations, uses enough energy to have the equivalent of 50 energy slaves. Americans are estimated to have 100 energy slaves apiece, potentially working 24 hours a day, 7 days a week. When an electronic device pesters us to update antivirus software, or download, install, and configure some program said to be improved, we count the cost in the currency of annoyance. Annoyance is something we all understand. But while we spend a few minutes each day being annoyed, those slaves continue to work, constantly consuming energy. Those few minutes of annoyance are some fraction of an energy slave's daily cost. You need energy – oil, coal, or natural gas, most likely – to experience your daily electronic annoyance. The cost of complexity is ultimately energy.

Indeed life has not always been like this. Our societies were once not only much simpler, they were also less costly. People got by on less energy and, we like to imagine, less annoyance. Today the price of modern life includes polluted beaches and waters and lives lost on the Deepwater Horizon. The moments of annoyance with electronic gadgets are possible because of the energy required to power them, mainly in the form of fossil fuels. We acknowledge the practitioners of slow living, but we suspect that their approach will not be widely adopted. It will not be adopted because it cannot be. We are stuck with our complexity and with what it costs. In this chapter we explain how this situation came about, and why we need to understand the energy–complexity spiral. There is a chain that connects how and why modern societies emerged to our dependence on fossil fuels, and to the consequences of employing those energy slaves.

The Nature of Cultural Complexity

What does it mean to be complex? Unfortunately the term has many technical meanings together with its meanings in everyday usage. How many concepts of complexity are there? To give some daunting examples: there are computational complexity (the resources required to execute algorithms) and algorithmic complexity (the shortest binary program that outputs a string). Computational complexity can be linear, logarithmic, or exponential. There are complex systems, complex mechanisms, and complex behavior. Complexity can be specified, irreducible, or unruly. To some ecologists, complexity is equivalent to diversity, whereas to others complexity emerges through hierarchy. Complexity can be hierarchical or heterarchical. Hierarchies can be simple and short, or complex and elaborate. Complexity can occur within a system, or by embedding types of subsystems. Organization in complex systems emerges through constraints, which can be asymmetrical or symmetrical. Added to this Tower of Babel, complexity means something akin to complicated or confusing in common usage. In the early days of the Santa Fe Institute (a research institute founded to study complexity), people used to say, "Complexity is perplexity." Adding to the confusion, people often use the term without specifying what they intend it to mean.

Fortunately we can cut through the confusion. There is a simple understanding of complexity that can guide us. When people complain about the complexity of modern life, they usually mean that they have too many things to do and not enough time to do them. It is important to understand this clearly. The difficulty arising from the complexity of modern life is not just having too much to do, although that is important, but having too many different kinds of activities that take up one's time. A person who performs a limited number of activities much of the time learns shortcuts and other ways to improve. This becomes harder as the activities diversify. Having too many different kinds of activities means, among other things, having to be in many different places, having to know many different kinds of things, and having to interact with various different technologies. This produces feelings of time stress, of having too much to do.

Consider a recognizable day for a suburban family of four with two professional parents. Early in the day the children must be awakened and fed their breakfasts. One child has early band practice, so dad leaves early to drive the child to school. This means that, even if he wants to, dad cannot use public transportation. The other child starts at the normal time and can, fortunately, take the school bus. Dad teaches at a local college and, when he starts his work computer, his day becomes more complex. The department chair wants to have a faculty meeting later in the day. Various programs on his computer nag him to download or install updates. A message from the college's computing services warns of a new type of online scam being sent through emails. Another message warns him that his old email program, which still meets his needs, will soon no longer function. He must get the new version. This new version has a different interface that he must master and for a while he cannot do his work. Students come by with academic or personal problems, the administration announces a new accounting program that he must be trained to use, and grades must be posted. Among all these things he must prepare lectures, read professional literature, do research, and write papers for publication.

Mom is a manager at a local business. Her day is much like dad's, with meetings, software updates, customer complaints, intraoffice personality conflicts, and no time for lunch. She often feels that she spends all of her time swatting at gnats, and cannot plan long-term strategy for the company. She, too, cannot manage to use public transportation, even though often she feels that she should. During the day mom and dad talk on their cellular telephones (which yesterday required them to set up new voicemail accounts) to organize the rest of the day.

There is no respite at the end of the work and school days. One child goes to soccer practice, the other to a dance lesson. The parents must drive them to both. One parent takes a car for an oil change during soccer practice and is late returning. The other goes shopping for children's clothing. Dinner is bought on the run and eaten in front of the television. Then the children must be supervised while doing their homework, and one needs special help with math. There are also bills to be paid, an evening chore. The other parent goes to a parent–teacher conference. The family computer comes on, and its software begins to download updates, including for the parental control software that mom and dad have had to learn. Eventually the children go to bed, and the parents sit to watch some television. Happily the cable connection is working tonight. But there are repeated alerts for a missing child. Mom and dad understand the need for the alerts, but the alerts make it impossible to relax and enjoy the television show. The parents give up and go to bed. Early tomorrow it starts all over again.

Life was not always like this. For comparison, let us look at an average day of a family in ancient Rome. Much in their lives would be like the lives of most urban residents until the late nineteenth century. We will make them a moderately wealthy family, wealthy enough to own their own home. It is a modest home, but that they own it is unusual. Ancient Rome had only about 1,800 such houses, compared to more than 42,000 multistory apartment buildings. We choose such an unrepresentative family because, like the Americans we just profiled, most of their work is done by slaves. Unlike the American energy slaves, the Roman ones were, of course, real humans.

The day begins with the slaves rising before dawn. The master and mistress (in separate bedrooms that we would consider tiny) are awakened at dawn. The master is shaved and plucked (both painful), while the mistress has make-up

Fig. 5.1 A street of Roman shops, Trajan's market, Rome

applied and her hair done. Breakfast is a hearty meal: bread, cheese, milk, fruit, and perhaps some meat. It might include leftovers from last night's dinner. After breakfast, the master goes to his office at the front of the house, where he receives clients. These are poorer people who come to him for a favor, a job, or some money. In return, the clients will support him when he runs for political office. He confers briefly with the administrator of an apartment building the family owns, and reviews accounts for their farm.

While the mistress supervises the slaves in shopping, cleaning, cooking, gardening, and teaching the children, the master walks to the forum to see his banker and look into business opportunities. He meets and talks with several acquaintances, always cultivating those who might be able to help him. This must all be done in the morning. The workday ends at noon and shops close for the day (Fig. 5.1).

The master returns home briefly. He and the mistress head for the baths, accompanied by two slaves. They get a quick lunch along the way, a small meal in the Roman day (Fig. 5.2). Most of the afternoon is spent at the baths, where one can exercise, bathe, and get a massage (Fig. 5.3). Most important, time at the baths is spent socializing, which includes cultivating more business and political opportunities.

Fig. 5.2 A Roman lunch counter in Ostia

Fig. 5.3 Roman baths at Pompeii

Dinner begins in late afternoon. Guests have been invited to a small banquet, which will go on for several hours. The guests are, of course, important people who can help the master's political career. Dinner ends before dark so the guests can return home safely. Because the only light comes from oil lamps, the master and mistress retire shortly after dark. The slaves work a little longer, but much of the cleanup from dinner will wait for the next day.

Life for a wealthy Roman sounds like the ideal of slow living, without the stresses and exasperations experienced by Americans. Our wealthy Romans live better lives, of course, than those who are poorer, but even merchants close their shops at noon and wile away the afternoon at the baths. Of course their lives are less comfortable than what we are accustomed to today. Houses of even the wealthy were dank and drafty, hot in the summer and cold in the winter. Sanitation was well developed, with plentiful fresh water, but ancient Rome's air was badly polluted. The Roman family we have sketched lives as well as they do because of the slaves they can afford. Having slaves frees the master to pursue a life in politics, the mistress to indulge in lengthy treatments for her face and hair, the children to be taught Greek, geometry, and rhetoric, and both parents to spend a relaxing afternoon at the baths nearly every day. Yet they have only a half-dozen human slaves, compared to 100 or more energy slaves at the command of even Americans of modest means. This is 400 energy slaves for an American family of four, 67 times what the Roman family commands. With so much energy at their command, why do the Americans lead such harried lives? Do they not buy automobiles, dishwashers, computers, and cellular phones to make life easier? In actuality, the availability of so much energy congests peoples' lives by complexifying the things that they must do.

The representative day for an average suburban couple, and the difference from a Roman day, illustrates some of what we mean by complexity, and helps to explain it. Complexity in this example consists of two elements: the diversity of things to be done, and the coordination required to get them done. These two elements introduce what the development of complexity in human societies has entailed. It has consisted of adding more parts, especially more kinds of parts, and organizing to coordinate those parts. In an early attempt to illustrate this point, the anthropologist Julian Steward contrasted small-scale hunting and gathering (foraging) societies with a small part of twentieth-century America. Most foragers studied by anthropologists lived in small groups, sometimes just individual families. They might coalesce into larger groups at certain times of the year when there was a temporary abundance of food in one place. The San (Bushmen) people of the Kalahari Desert in Botswana, for example, come together into larger groups when there is an abundance of mongongo nuts. They settle into a grove of mongongo trees and, literally, eat their way outward. For a few weeks, mongongo nuts are enough to power the society. Eventually the supplies in one place are exhausted, and small groups move on to look for food elsewhere. Sometimes, though, the large groups break up because of arguments. This, too, has a complexity lesson, which we take up below.

The Shoshone of the Great Basin also lived as mobile foragers, coming together when there was a large crop of pinyon pine nuts. These could be stored easily to support people through the winter. Anthropologists of the early twentieth century studied the Shoshone and other hunter-gathers of western North America to record the elements or traits of their cultures. These studies were done by some of the important figures in early American anthropology: Alfred Kroeber, Julian Steward, and John Peabody Harrington. Their approach today seems quaint. Cultures were broken down into discrete customs or elements, such as forms of descent (matrilineal, etc.), where a couple resides after marriage, types of houses or tools, and various kinds of cultural practices. The purpose was to list all of the discrete elements comprising a culture, and compare these among several cultures. Long ago anthropologists stopped viewing cultures this way, as if all the parts of a culture were commensurate elements. The Culture Element Distribution studies (as they were known) did, however, have a useful, unintended result. Julian Steward realized that culture element lists were a crude way to begin to compare societies on a dimension of complexity. To this end, he noted the contrast between the 3,000 to 6,000 culture elements documented among the native people of western North America, and the 500,000 artifact types that accompanied U.S. forces landing at Casablanca in 1942. Of course, the complexity of a military landing is but part of the complexity of the military as a whole, which was in turn only a part of the overall complexity of twentieth-century America. That was part of Steward's point.

There is a technical term for having many kinds of parts: structural differentiation. Structural differentiation is one aspect of complexity, but it is not sufficient to make a society complex. Something must make the parts work together, and that is organization. Organization is a subtler concept and worth some discussion. Here is a simple example of what the lack of organization means, and by implication, how it functions. Bushman society lacks institutional ranking (in which some people have regular power over others), other than the simple universals of a person's age and sex. Without authority figures, Bushmen cannot resolve disputes. They lack the organization that a hierarchy provides, and this means that there are things that they cannot accomplish. When they have an argument, sometimes the only solution is to move away. Organization in the form of institutionalized power would solve this problem.

The fact that complex systems need organization sheds a different light on the North African landings during World War II. The challenge of the logistical train that headed for the coast of northwest Africa was not just the great diversity in artifact types, but also how they were packed aboard ship. In proper combat loading, matériel should be stowed in reverse order from the sequence needed upon landing. No doubt the U.S. military understood this in principle, but failed to apply it in practice. Matériel cascaded into the departure docks chaotically, and was loaded onto ships haphazardly. Soldiers then broke the windshields of stowed vehicles that they had to climb over to find items. Explosives wound up in passageways, staterooms, and troop holds. Needless to say, unloading under fire was chaotic. Guns arrived on the beach without gunsights, without ammunition, without gunners. Important radio equipment had been stored as ballast because it was heavy. Medical supplies remained on the ships for 36 critical hours. To find any specific thing it was necessary to unload nearly everything.

The problem with the 500,000 artifact types shipped to Casablanca is that they did not, in fact, comprise a complex system. The system lacked organization. Differentiation in structure without corresponding organization makes a system complicated, not complex. In a complex system, certain parts constrain others, making the behavior of the constrained parts simple and predictable. Organization emerges from these constraints. Constraints can be active and overt (a superior's instructions, for example), or implicit (such as a spouse's raised eyebrows). The essence of the evolution of social and cultural complexity is, then, differentiation in structure combined with organization that increases to constrain the structure. As a human system becomes elaborate and organized, the behavior of its individual elements – institutions, social roles, or just individual people – is channeled and simplified. Where once in human societies, for example, most people could perform most tasks, today specialization is the norm. With specialization, behavior is constrained and predictable.

Conversely, though, organization alone cannot make a system complex. At the opening of the Beijing Olympics in 2008, we saw an impressive display

Fig. 5.4 Opening ceremony at the Beijing Olympics. (Source: Wikimedia)

of the organization of which China is capable. One display showed a sea of drummers, each dressed identically, all drumming in unison. It was most impressive, as it was meant to be (Fig. 5.4). But it was not a complex system. Although organization was high (everyone performed as required), there was no structural differentiation at all.

So complexity as used here refers to proliferation of structure (more parts, more kinds of parts, more kinds of things to do) and to the organization that is needed to make the parts and activities work together. Complex societies have great numbers of parts and high levels of organization. The proliferation of parts and activities can be seen today in such areas as technology (with many technologies themselves made up of numerous kinds of parts), in our institutions (think of all the acronyms of agencies in government), in the myriad jobs that we do, and in the diverse daily activities that cause so much stress. The organization to make these parts work together (which they do not always do) comes in the form of social norms, beliefs, rules, requirements, regulations, laws, instruction manuals, and so forth, all the things that make people behave in a predictable way. Today, more and more organization comes through electronic media.

Energy and the Development of Complexity

The most important point to understand in the emergence of complexity is that it is not free. Complexity costs. In the realm of complex systems there is, one might say, no free lunch. Whenever you add more parts to a system, there is a cost to those parts. Whenever you engage in more activities, there is a cost. The organization to integrate parts and activities also has costs. This is the case in all living systems, not just human societies. Consider the evolution of life from the earliest single-cell organisms to today's mammals. Mammals have many specialized parts for such things as movement, sight, hearing, smell, ingesting and digesting food, excreting waste, procreation, and so forth. There is a metabolic cost to each of these parts. The organism must consume more food to sustain the parts. A more complex organism is a more costly one.

It is the same for societies. A more complex society is more costly. Insect societies illustrate this clearly, and the lesson we can learn from them is direct and useful. Throughout the lowlands from southern Texas to South America there are species of ants that live in colonies and farm. They grow fungi on various organic materials that they bring into their nests. Fungus-farming ants emerged about 50 million years ago, and today there are at least 200 species of them. Some of these species illustrate important things about complexity and energy. At one extreme there are the ants of the genus *Myrmicocrypta*, which live in small colonies averaging about 100 insects. Their fungi grow on various bits of organic materials that they collect and bring into the nest, especially insect droppings. They forage close to the nest, out to about 1–1.5 meters distance. *Myrmicocrypta* societies are very simple: all individual ants are the same size and shape, and so there is little structural differentiation.

At the other extreme are ants of the genus *Atta*. These are the classic leaf cutters, whose trails through tropical forests are justly famous. Leaves are the substrate on which these ants grow their fungus, and they need very large quantities. *Atta* trails extend up to 100 meters with only a short distance between individual insects. These trails are among the most impressive visual features of tropical forests. Stand next to one, then look in one direction and the other. The trail extends as far as one can see. These trails are a reflection of *Atta* societies, which are large and highly organized. Colonies contain up to millions of individual insects. Their social structure is highly differentiated with many specialists: queen, soldiers, foragers, leaf cutters, and gardeners. Their bodies are specialized to their jobs with the largest *Atta* ants about 300 times bigger than the smallest. *Atta* are the most complex of ant societies.

Nitrogen is the prime resource for fungus-farming ants. The fungi require it. Nitrogen is provided by the items that the ants bring into their nests: droppings, insect bodies, flowers, and leaves. How much the colonies need depends on their size and complexity, and on the quality of the raw material. The small simple colonies of *Myrmicocrypta* haul about 40–152 grams of substrate material per month, whereas *Atta* colonies haul on average 76.5 kilograms of leaf fragments. The larger and more complex colonies need from 500 to 1,900 times as much resource per month as the simpler ones. This is why *Atta* ants form long lines of thousands of insects carrying leaves. They need such a large amount of substrate that it must be transported with high organization to achieve efficiency.

We encounter fungus-farming ants again later in the book. They have more to teach us about complexity and energy, and about energy and society. They may even help us catch a glimpse of what our future energy use may entail. For now, we emphasize the connection between greater complexity and a need for more energy.

It should start to become clear why modern societies use so much energy. We are complex: highly differentiated and highly organized. We tend to think of energy consumption as a matter of lifestyle. This is partly true, but a total picture is more subtle: complexity *requires* energy. Our example of the simplicity of life in ancient Rome is useful here again. The description came from a recent book titled *A Day in the Life of Ancient Rome*, by the Italian science journalist Alberto Angela. Angela points out how the handful of slaves facilitating life in an ancient household has been transformed into the 100 energy slaves we employ today. A bottle of gasoline provides energy equivalent to about 50 slaves pushing a small modern car for 2 hours. Electrical outlets in a home provide the labor equivalent of 30 slaves. Washing machines, stoves, microwaves, toasters, blenders, mixers, faucets, refrigerators, dishwashers, vacuum cleaners, water heaters, light bulbs, central heating, and so forth have all replaced the work previously done by slaves at a typical energy cost perhaps 25 times greater for the same ultimate result (e.g., cooked food, clean clothes). Beyond the basics of existence, our lives today involve institutions, technologies, and activities that ancient Romans never dreamed of. The total of these institutions, technologies, and activities introduces complexity that requires energy. Our energy consumption, we can see, comes from a combination of ease of lifestyle (provided by household appliances), activities that we find enjoyable (e.g., travel for pleasure), and complexity (more parts to maintain and more things to do, in part because energy slaves take care of the basics of existence).

Part of the complexity in our lives comes from the fact that so many of our technologies are single purpose. Your oven does not clean the dishes, but in ancient Rome the same slave may have done both cooking and cleaning. Each of these single-purpose technologies has an energy cost. Each also has a personal cost: the time and money (i.e., energy) that it takes to install an appliance, learn to operate it, have it periodically repaired, and ultimately dispose of it. This is one consequence of structural differentiation.

A more complex human society is more costly, not just in absolute terms but also per capita. For insects or animals we can calculate this cost as calories expended per grams of resource gained. Among humans the costs of complexity may be harder to identify, for the costs may occur in any of several different forms. One form that we all recognize is money. If an institution (a private entity or a public one) establishes a new bureau or begins a new line of activity, complexity increases and there is a monetary cost. We can see this in the regulations and new government bureaus that are emerging in response to the financial crisis of 2008–2009. When we go from automobiles with conventional engines to hybrid cars with two engines, complexity increases and there is a monetary cost (and more to break). If we establish a new branch of learning (biotechnology, say) and publishers then start new journals and universities establish new departments, there is a monetary cost. If the military develops a new type of weapon, complexity increases and we have seen, over and over, that new weapons cost. We could continue, but readers can surely think of examples from their own experience.

The cost of complexity can be counted also in other currencies. Labor or work is one. Vaclav Smil points out that for foragers, gathering roots can give a net energy return of 30–40 times the labor that one puts in. Gathering roots must be one of the most productive ways to get food. In our complex societies today, is there any activity in which human labor could yield a return of 30 times the cost without using fossil or nuclear energy? How many people in our complex societies already feel that they work too much? Add more things to an office worker's day, and as the complexity of the job increases, so does the labor cost. In some ways labor and money are alternative currencies because most of us labor for a salary. Work, of course, takes energy.

Time is another cost of complexity. So is annoyance or stress, inasmuch as these often arise from feeling that there is not enough time for all the things that we must do. Increased security at airports, for example, is another thing to cope with: a complexification. It has costs. We pay those costs through taxes, through the ticket price, and through the time and annoyance of waiting in line, emptying our pockets, removing computers and personal effects, and sometimes submitting to the indignity of an intrusive search.

All of these costs are transformations of energy. As with animals and insects, energy is for humans the true cost of complexity. In 1984, Cutler Cleveland, Robert Costanza, Charles Hall, and Robert Kaufmann published in *Science* a now-classic paper titled "Energy in the U.S. Economy: A Biophysical Perspective." Looking at 100 years of time-series data, and 87 economic sectors over a shorter period, they showed that energy is the basis of the U.S. economy. Gross national product, labor productivity, and price levels all rose directly with energy use. More recently, James Brown of the University of New Mexico and his colleagues have shown that many quality-of-life indicators can be predicted by per capita energy consumption. These include physicians per 100,000 people, life expectancy, infant mortality, meat consumption, patents, Nobel prizes, and others. Although people in the industrialized world, and especially Americans, like to think that we earned our way of life through ingenuity and hard work, in fact our way of life depends on consuming inexpensive but high-quality energy. Without energy, ingenuity and hard work could not provide the quality of life that we now enjoy. After all, our ancestors were ingenious and worked hard, and this is the case with many people today in developing countries. Yet none has achieved our standard of living. When we hear the word "cost" we think of money. In fact, as Cleveland, Brown, and their colleagues showed, money is a transformation of energy. We pay for complexity with high-quality energy. Therefore, energy is the currency that ultimately matters.

The other currencies that we have discussed, those of labor, time, annoyance, stress, are also transformations of energy. Time is money, in a popular saying, and money comes from energy. Time is also energy. We use energy for everything temporal that we do, including sleeping. Elapsed time equals consumed energy, even if only for human metabolism (100 watts on a continuing basis, as noted in Chap. 3). We save time (or try to save time) by substituting the work of energy-consuming appliances. Wastage of time (i.e., energy) produces annoyance, and the perception that we have not enough time leads to feelings of stress. Thus, all of the currencies in which we count the cost of complexity are transformations of energy.

The fact that complexity costs presents us with a dilemma. In the days before fossil fuels, the extra cost of supporting a more complex society meant that people worked harder. The cost was paid in the currency of human effort. Naturally no one wants to work harder than necessary. So if complexity meant that our ancestors worked harder, one must wonder: why did human societies become more complex? That may sound like a question of interest primarily to academics, but in fact it has important implications for life today and for our future. The cost of complexity has always provided some resistance against its growth. Yet time and again we have overcome that resistance, first with human labor, then with animal labor and simple machines, and finally with fossil energy and highly capable machines. The Deepwater Horizon was a good example of the last in this list. Because we have repeatedly overcome this resistance to complexity, we must ask if we will continue to do so. Will our societies grow still more complex and require still more energy per capita? We turn now to that question. And of course the corollary question is: will we be able to obtain more energy per capita in the future?

Problem Solving and the Development of Complexity

Cultural complexity is deeply embedded in our contemporary self-image, and this influences how we perceive it. Colloquially we know it by a more common term, "civilization," which we believe we have achieved through the phenomenon called "progress." This progressivist view supposes that great complexity (i.e., civilization) is intentional, that it emerged merely through the inventiveness of our ancestors. As we have seen, though, complexity costs, and therefore inventiveness alone cannot explain why complexity grows. The things we invent often tend to grow more costly as they grow complex. (There are countervailing trends in manufacturing efficiency that give the false impression that complexity becomes less expensive. Without added complexity, however, gains in manufacturing efficiency would produce even lower prices. Manufacturing efficiency, in any case, requires inexpensive energy.) The development of complexity requires facilitating circumstances. What were those circumstances? Archeologists once thought they had the answer: deep in our past, they reasoned, the discovery of agriculture gave our ancestors surplus food and, concomitantly, free time to invent urbanism and the things that comprise "civilization": cities, artisans, priesthoods, kings, aristocracies, and all of the other features of early states. The eminent archeologist V. Gordon Childe, for example, once wrote "On the basis of the neolithic economy further advances could be made … in that farmers produced more than was needed for domestic consumption to support new classes … in secondary industry, trade, administration or the worship of gods." Agriculture (i.e., surplus energy), it was reasoned, facilitated the invention of complexity among our ancestors, and we have continued to invent complexity ever since.

The progressivist view postulates, therefore, that complexity develops largely because it can, and that the factor facilitating this is surplus energy. Energy precedes complexity and allows it to emerge. This argument seems plausible, but it is actually too simple. There are significant reasons to doubt whether surplus energy has actually driven the development of complexity.

One strand of thought that challenges progressivism emerged in the eighteenth and nineteenth centuries in the works of Wallace (1761), Malthus (1798), and Jevons (1865). Malthus, in his well-known book on population and resources, was stimulated by the work of Wallace, who argued that progress would undermine itself by filling the world with people. Malthus in turn influenced Jevons, whose work we discuss shortly. In this century, the economist Kenneth Boulding developed three theorems from Malthus's essay on population. He labeled them the dismal theorem, the utterly dismal theorem, and the moderately cheerful form of the dismal theorem. We need not explore them all. It is useful to examine the utterly dismal theorem briefly because it directly challenges the progressivist view by focusing on one of its central tenets: improvements in the efficiency of technology, and the prospects of such improvement for increasing human welfare indefinitely. Boulding wrote:

> Any technical *improvement* can only relieve misery for a while, for as long as misery is the only check on population, the improvement will enable population to grow, and will soon enable *more* people to live in misery than before. The final result of improvements, therefore, is to increase the equilibrium population, which is to increase the sum total of human misery (emphases in original; Foreword to Malthus's *Population: The First Essay*, 1959).

The implication of this strain of thought is that humans have rarely had surplus energy. To the people of wealthy countries today, this may come as a surprise. We have lived in a time of surplus energy, and so we do not realize how unusual our time is. In the long span of human history, though, energy surpluses have quickly been dissipated by growth in consumption. Because humans have rarely had surpluses, the availability of energy cannot be the primary driver of increasing complexity.

We come back to our original question: why does complexity grow in the face of resistance? At least part of the answer is that complexity is a basic problem-solving tool. Confronted with problems, we often respond by developing more complex technologies (e.g., hybrid cars), establishing new institutions (e.g., the Department of Homeland Security), adding specialists or bureaucratic levels to an institution, increasing organization or regulation, or gathering and processing more information. Although we usually prefer not to bear the cost of complexity, our problem-solving efforts are powerful complexity generators. All that is needed for growth of complexity is a problem that requires it. Problems continually arise, thus there is persistent pressure for complexity to increase.

The evolution of complexity as a problem-solving tool is illustrated by the response to the terrorist attacks of September 11, 2001. In the aftermath, steps taken to prevent future attacks focused on creating new government agencies, such as the Transportation Security Administration and the Department of Homeland Security, consolidating existing functions into some of the new agencies, and increasing control over realms of behavior from which a threat might arise. In other words, our first response was to complexify by diversifying structure and function, and to increase organization and control. The report of the government commission convened to investigate the attacks (colloquially called the 9/11 Commission) recommended steps to prevent future attacks. The recommended actions amount, in effect, to more complexity, requiring more costs in the form of resources, time, or annoyance.

This is the normal way that complexity grows. We perceive a problem and attempt to solve it, usually by implementing a new organization, new activities, or new technologies with greater complexity. The extra costs seem, at the time, to be necessary and reasonable. We usually do not consider the cumulative cost of complexity in problem solving, but that is a topic for the next chapter.

Complexity, we see, can be viewed as an economic function. Individuals, institutions, and societies invest in problem solving, undertaking costs and expecting benefits in return. Our adoption of complexity follows some basic economic principles. In problem-solving systems, simple and inexpensive solutions are adopted before more complex and expensive ones. This is known as first plucking the lowest fruit. In the history of human food-gathering and production, for example, efficient hunting and gathering (San Bushman foragers actually work only a few hours per week) gave way to more labor-intensive agriculture (farmers work quite hard, especially when they do not have fossil fuels), which has in some places been replaced by industrial agriculture that consumes more energy than it produces. We produce minerals and energy whenever possible from the most economical sources, going to more inaccessible sources only when the need arises. Drilling in deep water in the Gulf of Mexico and elsewhere illustrates this process. Our societies have changed from egalitarian relations (no institutionalized differences in power), economic reciprocity (balanced economic exchanges), ad hoc leadership (situational and short-term), and generalized roles to social and economic differentiation, specialization, inequality, and full-time leadership. These changes are the essence of complexity, and they increase the costliness of any society.

In progressivist thinking, surplus energy precedes and facilitates the development of complexity. Certainly this is sometimes true: there have been occasions when humans adopted energy sources of such great potential that, with further development and positive feedback, there followed great expansions in the numbers of humans and the wealth and complexity of societies. These occasions have, however, been so rare that we designate them with terms signifying a new era: the Agricultural Revolution and the Industrial Revolution. We are in such a period now, enjoying the fruits of surplus energy.

Most of the time, cultural complexity increases from day-to-day efforts to solve problems. Complexity that emerges in this way will usually appear before there is additional energy to support it. Rather than following the availability of energy, cultural complexity often precedes it. Complexity thus compels increases in resource production.

The fact that complexity and costliness increase through mundane problem solving suggests a conclusion that some readers will find disturbing: contrary to what is often suggested in debates about energy, climate, and our future, *it is usually not possible for a society to reduce its consumption of resources voluntarily over the long term*. To the contrary, as problems great and small inevitably arise, addressing these problems requires complexity and resource consumption to increase. As presented in the next chapter, we know of only one case in history when a large complex society survived by simplifying and reducing its consumption of resources. The usual approach to solving problems goes in the opposite direction. To believe that we can voluntarily survive over the long term on less energy per capita is to assume that the future will present no problems. This would clearly be a foolish assumption, and the reality places one of the favorite concepts of modern economists and technologists, sustainable development, in grave doubt.

We are faced, then, with the prospect that under current trends, our societies will continue to grow in complexity, and in the energy that we will need. We must, at this point, consider an important question: could innovation reduce the energy cost of complexity? With more efficient technologies, could we continue to grow more complex and solve more problems without requiring more energy per capita? It is worth spending some time looking into this question.

Potential for Efficiency

In our technologically creative society, we place great faith in innovation. In the United States, creativity and innovation form a large part of the stories that we tell about our history. Alexander Graham Bell, Thomas Edison, and Henry Ford are among the pantheon of American heroes. We have to this point achieved so much innovation that we assume we will be able to rely on it in the future. In particular, we assume that any future shortage of resources, including energy, will be solved by innovations that improve technical efficiency, or we will develop new resources. In this view, we will be able to power automobiles for as long as we can improve miles per gallon. The current popularity of hybrid vehicles expresses this faith in technical innovation.

Our faith in innovation is enshrined in the pronouncements of both politicians and scholars. The first chapter has such a statement by Steven Chu, currently the U.S. Secretary of Energy. Secretary Chu's statement continues a long tradition of confidence in innovation. Here are some representative statements:

> No society can escape the general limits of its resources, but no innovative society need accept Malthusian diminishing returns. (Harold Barnett and Chandler Morse, *Scarcity and Growth: The Economics of Natural Resource Availability*, 1963)

> All observers of energy seem to agree that various energy alternatives are virtually inexhaustible. (Richard Gordon, *An Economic Analysis of World Energy Problems*, 1981)

> By allocation of resources to research and development (R&D), we may deny the Malthusian hypothesis and prevent the conclusion of the doomsday models. (Ryuzo Sato and Gilbert S. Suzawa, *Research and Productivity: Endogenous Technical Change*, 1983)

Based on this faith, many economists believe that energy and resources need not be considered in economic models. Resources are never scarce, they assert, just priced wrong. As a resource becomes harder to obtain, these economists believe, prices will rise and markets will signal that there are rewards to innovation. Responding to such signals, entrepreneurs will discover new resources, or develop more efficient ways of using the old ones. All it takes are incentives to do so. This belief is known as technological optimism. Clearly it is worth exploring this belief in some detail, for it is fundamental to questions about complexity, energy, and our future. If we can counter the cost of increasing complexity by becoming more efficient, perhaps this book is unnecessary. In Chap. 3 we suggested that the productivity of our system of innovation may actually be in decline. Now, to evaluate further the possibility of continual technological improvements, we need to understand how scientific disciplines develop.

There are many assumptions underlying technological optimism, one being that markets always work with perfect efficiency as long as there are no government distortions. The financial crisis of 2008–2009 has caused many people to question this assumption. We report here a different line of reasoning: technological optimists ignore complexity. Innovation is like any living system, human or otherwise. It grows in complexity and is subject to the benefits and costs that this imposes.

Institutionalized innovation as we know it today is a recent development. Our ancestors experienced periods of centuries to millennia with little or no technological change. In the Paleolithic (Old Stone Age, from the emergence of human ancestors to about 10,000 B.C.) there were even periods of technological stasis lasting hundreds of thousands of years. This is the statistically normal condition of human inventiveness. Innovation as we practice it today is an anomaly.

Innovation was rare in past societies in part because scientists were rare. As Derek de Solla Price suggested, "Society almost dared [scientists] to exist," throughout much of history. From the time of the ancient world through the eighteenth century, scholars and scientists were wealthy and self-sufficient, supported by students (as were ancient Greek philosophers) or by wealthy patrons, or were religious practitioners (such as Egyptian priests or medieval monks) who had time for inquiry. Toward the end of this period, the gentleman-scholar or -naturalist (or gentlewoman-scholar, such as Marie Curie-Skłodowska) emerged in the eighteenth and nineteenth centuries. The gentleman-naturalist (and his variant, the mad scientist) is an image that persists to this day in the public understanding of science, although it has long been quaint.

Today only a minority of research is done by an individual scientist in a white lab coat, working long into the night on some quixotic idea. Research today is mostly done by interdisciplinary teams. The reason for this development is that the early naturalists made themselves obsolete by depleting the stock of research quandaries that were relatively easy to answer. As the simplest research questions are answered, those next in line are more difficult and require the attention of diverse research teams. This is a normal and unavoidable process.

In every scientific and technical field, early research plucks the lowest fruit: the questions that are easiest to answer and most broadly useful. The principles of gravity and natural selection no longer wait to be discovered. Garvin McCain and Erwin Segal expressed this best. Science, they observed, is not likely to be advanced much farther by flying a kite in a thunderstorm or peering through a homemade microscope. As general knowledge is established early in the history of a discipline, the knowledge that remains to be developed is more specialized. Specialized questions become more costly and difficult to resolve. Research organization moves from isolated scientists who do all aspects of a project (the gentleman-naturalist), to teams of scientists, technicians, and support staff who require specialized equipment, costly institutions, administrators, and accountants. As one outcome of this process, the average number of contributors to scientific papers has been increasing. This is because research now requires the integration of more scientists who each specialize in some part of the whole. Thus fields of scientific research follow a characteristic developmental pattern, from general to specialized; from wealthy dilettantes and gentleman-scholars to large teams with staff and supporting institutions; from knowledge that is generalized and widely useful to research that is specialized and narrowly useful; from simple to complex; and from low to high societal costs.

As this evolutionary pattern unfolds, more resources and training are needed to innovate. In the first few decades of its existence, for example, the United States gave patents primarily to inventors with minimal formal education but much hands-on experience. After the Civil War (1861–1865), however, as technology grew more complex and capital intensive, patents were given more and more frequently to college-educated individuals. For inventors born between 1820 and 1839, only 8% of patents were filed by persons with formal technical qualifications. For those born between 1860 and 1885, 37% of inventors were technically qualified. As innovation grows harder, it takes more education and training to become a successful inventor.

It has long been known that within individual technical sectors, the productivity of innovation declines over time. In 1945, Hornell Hart showed that innovation in specific technologies follows a logistic curve: patenting rises slowly at first, then more rapidly, and finally declines. The great physicist Max Planck thought that science as a whole would experience diminishing productivity as it grew and exhausted the stock of things that are easy to learn. The philosopher Nicholas Rescher, paraphrasing Planck, observed that "… with every advance [in science] the difficulty of the task is increased." Writing specifically in reference to natural science, Rescher suggested:

> Once all of the findings at a given state-of-the-art level of investigative technology have been realized, one must move to a more expensive level …. In natural science we are involved in a technological arms race: with every "victory over nature" the difficulty of achieving the breakthroughs which lie ahead is increased (*Unpopular Essays on Technological Progress*, 1980).

In tribute to the famous physicist, Rescher termed this "Planck's Principle of Increasing Effort." Planck and Rescher suggest that exponential growth in the size and costliness of science is needed just to maintain a constant rate of innovation. Science must therefore consume an ever-larger share of national resources in both money and personnel. Jacob Schmookler, for example, showed that although the number of industrial research personnel increased 5.6 times from 1930 to 1954, the number of corporate patents over roughly the same period increased by only 23%. Such figures prompted Dael Wolfle in 1960 to write an editorial for *Science* titled "How Much Research for a Dollar?" Derek de Solla Price observed in the early 1960s that science even then was growing faster than both the population and the economy and that, of all scientists who had ever lived, 80–90% were still alive at the time of his writing. At the time of our own writing, there are discussions of boosting the productivity of American science by doubling the budget of the National Science Foundation, just as the research budget of the National Institutes of Health was doubled a few years ago.

The stories that we tell about our future assume that innovation will allow us to continue our way of life in the face of climate change, resource depletion, and other major problems. The possibility that innovation overall may produce diminishing productivity calls this future into question. As Price pointed out, continually increasing the allocation of personnel to research and development cannot continue forever or the day will come when we must all be scientists.

In 2005, Jonathan Huebner published an article with the provocative title, "A Possible Declining Trend for Worldwide Innovation." Huebner is a physicist at the Naval Air Warfare Center in China Lake, California (although his innovation research was done independently). Using 7,200 major innovations listed in an important work, *The History of Science and Technology*, by Alexander Hellemans and Bryan Bunch, he plotted key innovations over time against population to investigate whether there is an economic limit to innovation. Looking at today's unending stream of inventions and new products, most people assume that innovation is accelerating. Ever-shorter product cycles would lead one to believe so. In fact, relative to population, innovation is not accelerating. It is not even holding steady. Huebner found that major innovations per billion people peaked in 1873 and have been declining ever since. Then, plotting U.S. patents granted per decade against population, he found that the peak of U.S. innovation came in 1915. It, too, has been declining since that date. Compare this observation to Fig. 3.16 in Chap. 3.

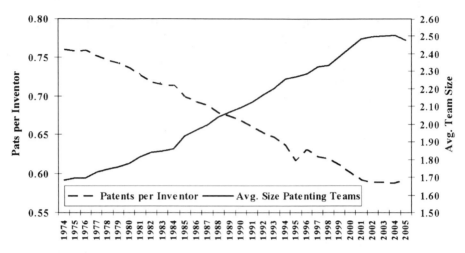

Fig. 5.5 Average size of patenting teams and patents per inventor, 1974–2005. Source: 18

Huebner's analysis produced some other startling facts. Although every year we are offered new or improved electronic gadgets, in fact key innovations in 2005 had dropped to the same rate that humanity achieved in 1600. Despite massive spending on research and education, it is harder today to make a fundamental breakthrough than it was 100 years ago. We are indeed, suggests Huebner, approaching an economic limit to innovation.

There have been criticisms of Huebner's work, particularly the selection of key innovations on which he relied. Recently Deborah Strumsky of the University of North Carolina and José Lobo of Arizona State University teamed with one of us (Tainter) in a systematic investigation of the productivity of innovation. Employing the very large database of the U.S. Patent and Trademark Office (USPTO), we investigated the productivity of innovation in a number of fundamental technical fields, including surgery and medical instruments, metalworking, optics, drugs and chemicals, energy technologies, information technologies, biotechnology, and nanotechnology. Our results are consistent with Huebner's general findings.

Our measure of productivity is patents per inventor. The USPTO has only recently begun to keep records that allow such an analysis. The results are illuminating. Figure 5.5 shows that from 1974 to 2005, the average size of a patenting team increased by 48%. This parallels the trend, noted above, toward increasing numbers of authors per scientific paper. The increasing numbers of authors in both invention and publication derive from the same source. This is the increasing complexity of the research enterprise, required

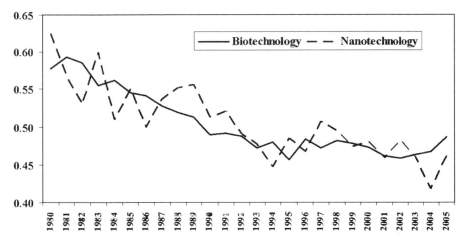

Fig. 5.6 Productivity of innovation in biotechnology and nanotechnology, 1980–2005.
Source: 18

to meet the increasing difficulty in the questions addressed or the breakthroughs sought, and leading to the incorporation of more and more specialties in an individual project.

The scientific enterprise has been growing larger and larger, but it is producing fewer and fewer innovations per inventor. Over the period shown in Fig. 5.5, again from 1974–2005, patents per inventor declined by 22%. We should emphasize that in a period of just over 30 years, the length of an average career, the productivity of innovation has declined by more than one-fifth. That is a finding of the highest importance for assessing our future.

As Hornell Hart showed in 1945, the characteristic evolution of a technology is logistic: innovations come slowly at first, then accelerate for a while, and finally come more slowly and with greater difficulty. This opens the possibility that higher productivity in newer technical fields might offset declines in older ones. To investigate this possibility, we produced the chart shown in Fig. 5.6. Even in the new fields of biotechnology and nanotechnology there is diminishing productivity of innovation. If this occurs even in new fields, then the problem is clearly intrinsic to science as a whole, and not limited to individual fields.

We have also investigated the productivity of innovation in the energy sector, as shown in Fig. 5.7. Here as in other technical fields, the productivity of innovation is declining. It is declining not only in older fossil fuel technologies, but also in the wind and solar technologies that many people hope will power our future.

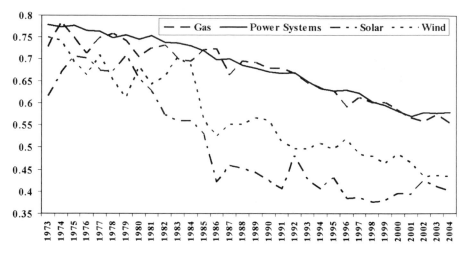

Fig. 5.7 Productivity of innovation in the energy sector, 1974–2005. Source: 18

The reason for the diminishing productivity of innovation is complexity. Scientific fields, as we have described, undergo a common evolutionary pattern. Early work establishes the boundaries of the discipline, sets out broad lines of research, establishes basic theories, and solves questions that are inexpensive but broadly applicable. Yet this early research carries the seeds of its own demise. As pioneering research depletes the stock of questions that are inexpensive to solve and broadly applicable, research must move to questions that are increasingly narrow and intractable. Research grows increasingly complex and costly as the enterprise expands from individuals to teams, as more specialties are needed, as more expensive laboratories and equipment are required, and as administrative overhead grows. We have an impression today that knowledge production continues undiminished. Each year sets new records in numbers of scientific papers published. Breakthroughs continue to be made and new products introduced. Yet we have this impression of continued progress not because science is as productive as ever, but because the size of the enterprise has grown so large. Research continues to succeed because we allocate more and more resources to it. In fact, the enterprise does not enjoy the same productivity as before. It is clear that to maintain the same output per inventor as we enjoyed in, say, the 1960s, we would need to allocate to research even greater shares of our resources than we do now. Without such an allocation, the productivity of research declines.

In 1963, Derek de Solla Price wrote that science could not continue to grow as it has over the past two centuries. He suggested that growth in science

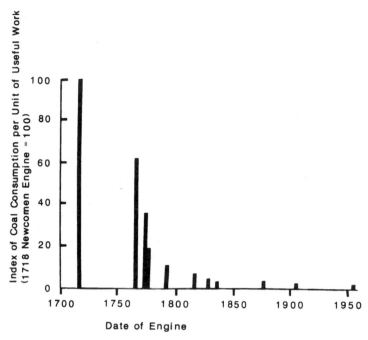

Fig. 5.8 Diminishing returns to improving the steam engine. (Source: Richard G. Wilkinson, 1973. *Poverty and Progress: An Ecological Model of Economic Development.* Methuen, London, p. 144)

could continue for less than another century. As of this writing, nearly half that time has elapsed.

This does not mean that there will be a quick end to improvements in technical efficiency in the energy-consuming machines on which we rely. For some time we surely will continue to experience such improvements. It seems likely, though, that such improvements will become harder and harder to achieve and that increments of improvement will become smaller and smaller. Consider the improvements to the steam engine, as shown in Fig. 5.8. Here the major improvements came with Watt's steam engine. Improvements thereafter became smaller and smaller as thermal efficiency increased. A doubling of efficiency in the twentieth century would save much less fuel per engine than a 10% increase in the eighteenth century, and the savings would be much harder to achieve. This is the typical evolutionary pattern of efficiency improvements.

Moreover, improvements in efficiency often produce paradoxical results. As we noted in Chap. 2, in 1865 the noted British economist William Stanley Jevons (1835–1882) published a now-classic work titled

The Coal Question. Jevons was concerned that Britain would lose its economic dynamism and pre-eminence in the world due to an inevitable depletion of its reserves of easily mined coal. Of course he did not foresee the dominance of petroleum, even denying its likelihood, and so the central worry of the book turned out to be misplaced. But *The Coal Question* contains a gem that enshrines the book as among the most significant works of resource economics. That gem is known today as the Jevons paradox. It cannot be expressed better than in Jevons' own Victorian prose.

> *It is wholly a confusion of ideas to suppose that the economical use of fuel is equivalent to a diminished consumption. The very contrary is the truth* [emphasis in original].
>
> As a rule, new modes of economy will lead to an increase of consumption ....
>
> Now, if the quantity of coal used in a blast-furnace, for instance, be diminished in comparison with the yield, the profits of the trade will increase, new capital will be attracted, the price of pig-iron will fall, but the demand for it increase; and eventually the greater number of furnaces will more than make up for the diminished consumption of each.

In short, as technological improvements increase the efficiency with which a resource is used, total consumption of that resource may increase rather than decrease. This paradox has implications of the highest importance for the energy future of industrialized nations. It suggests that efficiency, conservation, and technological improvement, the very things urged by those concerned for future energy supplies, may actually worsen our energy prospects.

Does this mean that efficiency improvements are not worthwhile? Of course not. Efficiency improvements are highly valuable, but their value has a limited lifespan. Technical improvements may merely establish the groundwork for greater resource consumption in the future. This in turn requires further technical innovation, but as we have just discussed, those technical improvements will become harder and harder to achieve. And as we do achieve them, they may serve us for shorter and shorter periods. We have a tendency to assume that technical innovations such as hybrid automobiles will solve our energy problems. This is unlikely. As seen in Fig. 5.9, as more fuel-efficient cars entered the U.S. fleet in the late 1970s, Americans did not pocket the money they could have saved. Instead they simply drove more miles. Technical improvements do buy us time, which is itself worthwhile and may be all we can expect. One conclusion is inescapable: given the irreversible pattern of declining productivity of innovation, the cost of complexity cannot be offset forever by improvements in efficiency.

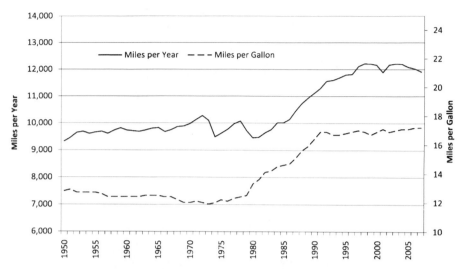

Fig. 5.9 The rebound effect: Fuel economy and annual miles driven, U.S., 1950–2007. (Source: http://www.eia.doe.gov/aer/txt/ptb0208.html. Accessed 4 Dec 2010)

The Energy–Complexity Spiral

Biologists understand that if you add energy to an ecosystem, that system will change. These changes may not be desirable. Add nitrogen or phosphorus to a lake, for example, and algae will proliferate, shutting out sunlight from lower depths. For our purposes here, the important point is that ecosystems cannot have surplus energy. If there is unused energy in a natural system, some species or combination of species will emerge to use it. That may involve expansion of an existing species, immigration of species from elsewhere, or the emergence of a new species. If you add extra energy to an ecosystem, it will complexify.

This is also the case in human societies. Give us extra inexpensive energy, and we will not leave it unused for long. Put some new form of energy into a human society, and that society will complexify. It will develop new activities, new technologies, new forms of entertainment and recreation, new institutions, new social roles, and it will develop and transmit new kinds of information. Each new activity, new institution, and so forth needs energy to continue, an ongoing energy budget. This process should sound familiar. It is what we have been doing since fossil fuels assumed such a prominent place in our lives. We found a new, abundant, inexpensive source of energy, and proceeded to use it and complexify.

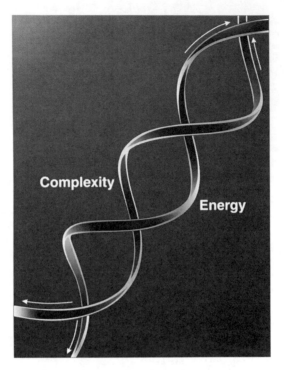

Fig. 5.10 The energy–complexity spiral

As noted earlier in this chapter, however, having surplus energy is a rare experience in human history. The fact that we are in such a special period leads many people to think that today's conditions are normal. In fact, they are not. Today's conditions of inexpensive energy are highly unusual, an aberration of history. We do not know how long this aberration will last, but it cannot last forever.

Most of the time, complexity increases not because it can, but because it is a good way to solve problems. We solve problems by developing more complex technologies, adding new institutions or social roles, or processing new types of information. Complexity that emerges to solve problems typically emerges before we have the energy that it requires. Complexity then compels us to increase the production of resources still further. In time that will mean using resources of lower quality that require increasingly complex technologies to find and extract.

This, then, is the energy–complexity spiral: complexity grows because we have extra energy, complexity grows because we must solve problems, and complexity requires that energy production increase still more (Fig. 5.10).

It is a spiral that we have lived with for two centuries or more. So far we have coped with this spiral fairly well. We have been able to increase the production of energy and other resources sufficiently to meet demand and to address the problems that we choose to solve. The challenge is that the petroleum which underpins much of this spiral is becoming harder and more costly to find and extract. As seen in Chap. 4, it is also requiring technologies of exploration and production that are growing more complex and risky. Failures of these technologies can, as we have seen, have catastrophic consequences. The fact that complexity grows to solve problems leads to the sobering realization that there may be no way out of this spiral, or at least no way that we would find desirable.

What happens to a society when it grows more and more complex on the basis of resources that become harder and harder to supply? Fortunately we need not guess. The human experience gives us examples that allow us to understand our current predicament, and to foresee how our future is likely to unfold. We explore those experiences in the next chapter.

Further Reading

Energy Slaves

1. http://www.earthinfo/page/Energy+slave. Accessed 17 Jan 2011

Energy and Cultural Complexity

2. Allen, T.F.H., Tainter, J.A., Hoekstra, T.W.: Supply-Side Sustainability. Columbia University Press, New York (2003)
3. Angela, A.: A Day in the Live of Ancient Rome, translated by Gregory Conti. Europe Editions, Milan (2009)
4. Boulding, K.E.: Foreward. In: Malthus, T.R. (ed.) Population: The First Essay, pp. v–xii. University of Michigan Press, Ann Arbor (1959)
5. Brown, J.H., Burnside, W.R., Davidson, A.N., DeLong, J.P., Dunn, W.C., Hamilton, M.J., Mercado-Silva, N., Nekola, J.C., Okie, J.G., Woodruff, W.H., Zuo, W.: Energetic limits to economic growth. Bioscience 61, 19–26 (2011)
6. Cleveland, C.J., Costanza, R., Hall, C.A.S., Kaufmann, R.: Energy and the U.S. economy: a biophysical perspective. Science 225, 890–897 (1984)
7. Jevons, W.S.: The Coal Question: An Inquiry Concerning the Progress of the Nation and the Probably Exhaustion of Our Coal-Mines, 2nd edn. Macmillan, London (1866)
8. Smil, V.: Energy in World History. Westview, Boulder (1994)

9. Tainter, J.A.: The Collapse of Complex Societies. Cambridge University Press, Cambridge (1988)
10. Tainter, J.A., Allen, T.F.H., Little, A., Hoekstra, T.W.: Resource transitions and energy gain: contexts of organization. Conserv. Ecol. 7(3), 4 (2003), http://www.consecol.org/vol7/iss3/art4

Innovation

11. Hart, H.: Logistic social trends. American Journal of Sociology 50, 337–352 (1945)
12. Huebner, J.: A possible declining trend for worldwide innovation. Technological Forecasting and Social Change72, 980–986 (2005)
13. Jones, B.F., Wuchty, S., Uzzi, B.: Multi-university research teams: shifting impact, geography, and stratification in science. Science 322, 1259–1262 (2008)
14. McCain, G., Segal, E.M.: The Game of Science, 2nd edn. Brooks/Cole, Monterey (1973)
15. Price, D. de Solla.: Little Science, Big Science. Columbia University Press, New York (1963)
16. Rescher, N.: Scientific Progress: A Philosophical Essay on the Economics of Research in Natural Science. University of Pittsburgh Press, Pittsburgh (1978)
17. Rescher, N.: Unpopular Essays on Technological Progress. University of Pittsburgh Press, Pittsburgh (1980)
18. Strumsky, D., Lobo, J., Tainter, J.A.: Complexity and the productivity of innovation. Systems Research and Behavioral Science 27, 496–509 (2010)
19. Wuchty, S., Jones, B.F., Uzzi, B.: The increasing dominance of teams in production of knowledge. Science 316, 1036–1039 (2007)

Chapter 6

The Benefits and Costs of Complexity

Here is how to boil a frog. Place the frog in a pan of tepid water. Raise the temperature so gradually that the frog does not realize it is being cooked. It may even fall into a stupor, as a person might in a hot bath. Eventually it will die. According to experiments done in the nineteenth century, you can indeed boil a frog this way. Biologists today claim that you can't. Either way, please don't try it.

Boiling a frog is a metaphor for the problem we all have perceiving changes that are gradual but cumulatively significant, that may creep up and have devastating consequences: a little increase here, a little there, then later some more. Nothing changes very much and things seem normal. Then one day the accumulation of changes causes the appearance of normality to disappear. Suddenly things have changed a great deal. The world is different, and it has been altered in a manner that may not be pleasant.

Sprinkle more and more sand on a pile and eventually you will produce an avalanche. The same thing happens in avalanches of snow. In 2000, Malcolm Gladwell published the book *The Tipping Point: How Little Things Can Make a Big Difference*. Tipping points come about when an accumulation of small changes suddenly produces something big, such as an avalanche. The French scholar René Thom developed a branch of mathematics to describe this, calling it catastrophe theory. What is a catastrophe? In Thom's mathematics it is simply a discontinuity. Despite the connotation of the title of the theory, in mathematics catastrophes are not intrinsically negative. They are simply points where an accumulation of small steps produces a sudden

J.A. Tainter and T.W. Patzek, *Drilling Down: The Gulf Oil Debacle and Our Energy Dilemma*, DOI 10.1007/978-1-4419-7677-2_6, © Springer Science+Business Media, LLC 2012

change of state, like tipping points. These points can be described with elegant equations and graphs, and classified into types of catastrophes.

The development of complexity follows this process. Encounter a problem and the response, as we discussed in the last chapter, is to adopt some solution that is more complex. This always seems reasonable. After all, one has a problem and it must be solved. Thus complexity grows, one step at a time. So does the cost of complexity. Costs increase by small increments, each reasonable and seemingly affordable—until, that is, the point of catastrophe is reached.

We know how to boil a frog. Complexification is how to boil a society. Complexity grows by small steps, each seemingly reasonable, each a solution to a genuine problem. We can afford the cost of each increment. It is the cumulative costs that do the damage, for the costs of solving previous problems have not gone away and now we are adding to them. The temperature increases insensibly and we are lulled into complacency. Eventually these costs drive a society into insolvency. A few people always foresee the outcome, and always they are ignored.

Complexity is not intrinsically good or bad. It is useful and affordable, or it is not. In this chapter we show how complexity can affect societies negatively, producing catastrophes that are real as well as mathematical. This is not, however, an inevitable outcome. There are ways to cope with complexity, to make certain that it serves us rather than conversely. One of the most important factors is the net benefit of complexity, and this changes over time. Another factor is how we pay for it. Various human societies have had different experiences with complexity. These experiences help us to understand our situation today, including why we need machines such as the Deepwater Horizon. The same experiences illustrate some of our options for the future. We explore these topics in this chapter.

The Net Benefits of Complexity

Complexity costs, as we know, but that is not all we need to know. It is equally important to understand the net benefits of complexity. As we discuss, the net benefits of complexity are so vital that they decide the fates of societies and civilizations. Net benefits will also interact with the supply of energy to determine our future.

In 1949, the Harvard linguist George Zipf published a book titled *Human Behavior and the Principle of Least Effort*. The Principle of Least Effort is intuitively obvious, although it can be hard to apply in practice. The principle holds that humans and animals try to accomplish their goals at the least cost. As discussed in Chap. 5, there are many ways to count costs, but they are all transformations of energy. The least-cost principle is obvious, but it does not always hold up in practice. We all know people who like to work, and who work more than they need to (although they may still be economizing given their motivations or goals). Many people will walk farther than necessary for the sake of exercise. Even if one wants to conserve effort, people do not always know the optimum or least-cost way to do things. As a general principle, though, least cost is generally valid. Even if we work more than necessary, few of us volunteer to pay more for our purchases than we feel we should. Even if we walk more than is required, we generally do not want to pay extra for our walking shoes. Colloquially, this is also known as taking the path of least resistance or plucking the lowest fruit first.

We employ the Principle of Least Effort when we look for the resources that we need. No one digs a mine for gold if it can be panned from a stream. We would never have considered looking for oil in deep water before we had fully developed the easy oil in Pennsylvania, Texas, Oklahoma, Louisiana, and California. We follow the same principle in the development of human society and in other aspects of history. As discussed in Chap. 5, and as widely known, human societies have developed from less to more complex. Following the Principle of Least Effort, the earliest human societies were as simple as they could be. This is also the case with the ant societies discussed in Chap. 5. Analyses of mitochrondial DNA show that simple societies of fungus-farming ants, such as those of the genus *Myrmicocrypta*, were the first to emerge. The large complex societies of leaf-cutter ants, such as those of the genus *Atta*, emerged much more recently. In their evolution, these ants followed the Principle of Least Effort.

In the evolution of complexity, the point of the Principle of Least Effort is that in human organization as in other matters, we first pluck the low-lying fruit. In the realm of complexity of organization, the low-lying fruit is a society of generalists (low structural differentiation), in which all men can do most of the tasks that other men do, and the same for women. Of course, some people have skills that others lack, and so may do more of certain things. But these differences are not institutionalized. Correspondingly there is low organization. That is, there are no institutionalized differences in power (other than those of age and sex), and so there is little ability to coerce others. This is the simplest and least costly society, and so it emerged first.

Because the least costly social institutions emerged first, societies have grown more costly as they have grown complex. But this does not mean that the benefits and costs of complexity change linearly. A linear relationship between these elements would mean that, as the benefits of complexity increase by one unit, for example, the costs of complexity always increase by one unit as well. In actuality, the relationship of the benefits of complexity to the costs is nonlinear. This simple fact is responsible for some of the major events of history. In the process of complexification, the relationship of benefits to costs is nonlinear because of the Principle of Least Effort.

If, in complexity as in other realms, we first pluck the low-lying fruit, this means that as complexity increases, societies always become more expensive. Many of our societies have been growing more complex and costly for the past 12,000 years or so. These days, thanks to the global expansion of industrialized nations, even societies that were once isolated and remote are today developing versions of first-world complexity. Occasionally this process is interrupted. There are times when societies collapse, that is, simplify rapidly. This has happened often enough that, as discussed below, we now have some understanding of why. Most of the time though, especially today, societies seem just to become more and more complex. A large part of the reason is that, as discussed in Chap. 5, complexity grows to solve problems, and there is never any shortage of those.

Complexity, then, is an investment. It has benefits and costs, and naturally we want the benefits to exceed the costs. As is generally true of economic processes, it is the initial investments in complexity that yield the highest returns. This is a variant of plucking the lowest fruit. The second fruit to pluck is the next one up, and so forth. In other words, when further complexity is required to solve problems, we next develop technologies and institutions that are just a bit more complex and costly. Whatever the problem is, it is solved at only a slightly higher cost.

Yet as we exhaust the least costly ways to organize society, make tools, obtain resources, and process information, only more costly solutions remain. Humans began by practicing the Principle of Least Effort, but after a while least effort no longer sufficed. The lowest fruit had been plucked. The graph in Fig. 6.1 is a device common in economics. Ordinarily such a graph shows the relationship of input to output, or costs to benefits. We can use it as a metaphor for the development of complexity in a society. Provided that people are averse to extra labor, the benefit–cost ratio of complexity will at first increase favorably. At the left side of the graph in Fig. 6.1, there are positive returns to complexity. Complexity is a worthwhile investment.

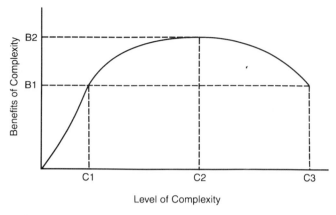

Fig. 6.1 Benefits and costs of complexity

At some point, however, the costs start to accelerate and the benefits of complexity, the ability to solve problems, increase more slowly. This is a normal economic event, and it is known as the point of diminishing returns. In Fig. 6.1, this point is shown as B1, C1.

This type of graph is referred to as a marginal product curve. Marginal product or marginal return is the extra output you get for an extra unit of investment. As an economic investment, complexity is subject to declining marginal returns.

Societies increase in complexity to solve problems, including the vital problems of their own existence. Sometimes societies increase in complexity just to maintain the status quo. Such endeavors are successful when the status quo is indeed maintained: the frontiers are defended, internal order has been restored, the financial system has been stabilized, or people again have jobs or enough to eat. If the costs and complexity of problem solving grow merely to restore a system to its previous condition, axiomatically the marginal return on investment in complexity declines. This is seen clearly in recent efforts of international intervention. In Somalia, in 1992, the United States and other nations deployed a complex military machine, voracious in its consumption of oil, merely to provide Somalis with stability and a daily food allowance of 2,500 calories, or so, per person. Some years before, Somalis had all these things without the cost of an international effort. For all the expenditures, a stable government could not be imposed from without, and to secure the delivery of food with helicopters yields nearly the lowest imaginable return on caloric expenditures. The result is that Somalia became a place that cost much more to sustain than ever before, yet its people were no better off than before its civil war. Much effort went to restore a status quo.

At the time of this writing (August 2010) the same process is underway in Pakistan. Large sections of the country have been flooded, and millions of people have lost their homes and livelihoods. Billions of dollars will be spent to aid Pakistan, and it will all go to restore Pakistan and its people to the life they had before this tragedy.

In Chap. 9, we explore in some detail the phenomenon of undertaking vast expenditures to maintain or restore a status quo. This process will have a profound influence on our future. We can see the process underway already. Consider the irony of all the petroleum-burning ships, planes, and helicopters operating to restore the Gulf to the condition that Nature gave it, freely and without our help, before the spill. We expend energy to get energy, then we expend more energy to reverse the damage caused by our efforts to get energy. Similar ironies are at work in our involvement in the Middle East. Regardless of what Western leaders claim, we would not be so concerned about having stable, peaceful governments in this region if it was the world capital of, say, banana production. Let us be honest: what we want from the Middle East is oil, and stable governments to provide it. The status quo in this case is a steady flow of petroleum. Our military is unmatched in its thirst for oil. So we send it to the Middle East, where it uses a lot of oil to secure the very same substance.

Referring again to Fig. 6.1, the decline in marginal returns to complexity begins to deflect more sharply beyond point B1, C1 on the curve. A society at this point becomes increasingly vulnerable to collapse. Two factors make a society more liable to collapse at this point. First, a society in this condition finds its fiscal strength depleted and productive capacity sapped as it spends more and more to accomplish proportionately less and less. When major adversities arise, as inevitably they do, there is no longer surplus economic capacity with which to confront them. If the crisis is survived, there may be less ability to meet the next one. Collapse thus becomes a matter of mathematical probability. In time an insurmountable challenge will come along. To anticipate a historical example that we discuss below, if the Roman Empire had not been toppled by Germanic tribes, it would have been later by Arabs or Mongols or Turks.

Second, declining marginal returns make complexity a less attractive strategy. The extraction of resources needed to sustain such a course of complexification, typically through taxes, alienates more and more of the population. People may not be able to express their dissatisfaction in terms of marginal productivity, but they can perceive when the taxes they pay do not yield a satisfactory return. Exit the Cold War and enter rising numbers of separatists, militias, and tax resisters.

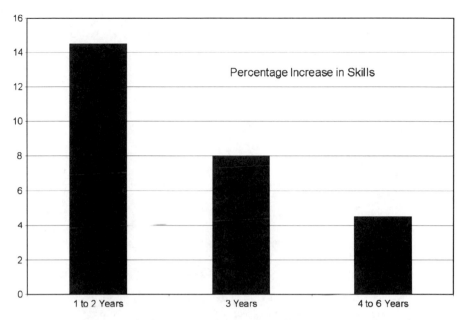

Fig. 6.2 Productivity of educating workers in the early Soviet Union. (Data from L. I. Tul'chinskii, 1967. Problems in the profitability of investments in public education. *Soviet Review* 8(1): 46–54)

Is there evidence that societies do in fact evolve this way? Indeed there is. First, consider education, which most of us agree is a good and useful thing. As a society increases in complexity and becomes more dependent on information and also more competitive, workers require higher levels of education. In 1924, S. G. Strumilin assessed the productivity of education in the nascent Soviet Union. The first 2 years of education, Strumilin found, raised a worker's skills an average of 14.5% per year. The productivity of education declined by adding a third year, for this raised the worker's skills only an additional 8%. Four to six years of education raised skills only a further 4–5% per year (Fig. 6.2). Clearly in this case, there were diminishing returns to additional education, at least on average.

In Chap. 5 we discussed an important example of diminishing returns to knowledge production. As easy research questions are answered and simple products introduced, research and development grow more complex and costly. The cost of innovation rises and the productivity of the effort declines (Figs. 5.5–5.7). Allocating more and more resources to research and development produces diminishing returns.

Our system of health care shows declining marginal productivity as it grows ever more complex. In 1930 the United States spent 3.3% of its Gross

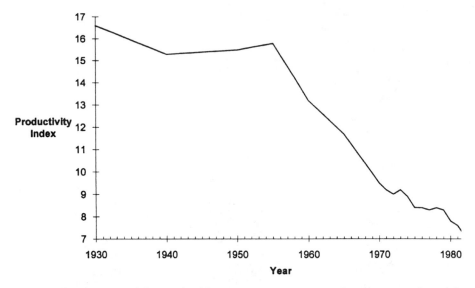

Fig. 6.3 Productivity of the U.S. health care system, 1930–1982. Productivity index = life expectancy/national health expenditures as percent of gross national product. (Sources: (1) Nancy L. Worthington, 1975. National health expenditures, 1929–74. *Social Security Bulletin* 38(2): 3–20. (2) U.S. Bureau of the Census, 1983. *Statistical Abstract of the United States: 1984* (104th edition). U.S. Government Printing Office, Washington, DC)

National Product (GNP) to produce an average life expectancy of 59.7 years. By 1982 medical expenditures had grown to 10.5% of GNP. This is a telling figure in itself: when one field of social investment takes an increasing share of a nation's wealth, the share available to all other sectors, including energy production, obviously declines. The growth in medicine's share of national wealth produced in 1982 a life expectancy of 74.5 years. This is a worthy figure, yet it represents a 57% decline in the productivity of our medical investments over a period of 53 years (Fig. 6.3). If our investment in medicine in 1982 had been as productive as it was in 1930, average life expectancy would have risen to 190 years. As it is, each extra year of national life expectancy is bought at a cost of lessening our prospects in other areas, because there is proportionately less to invest in education, infrastructure, or other kinds of research.

If complexity develops by the Principle of Least Effort, and therefore produces diminishing returns, how did our societies come to be as they are today? How did we become so prosperous? That is where those energy slaves come in. One way to pay for complexity is to find some subsidy to bear the cost. This is what we have done with fossil fuels. With energy slaves to do the work, diminishing returns are a less burdensome problem.

How did this come about? Richard Wilkinson showed in his 1973 book *Poverty and Progress* that it was by necessity. From the fourteenth through the eighteenth centuries the population of England grew, and as it grew, cut its forests. Soon there was a shortage of wood, which people needed every day. Coal came to be used in its place, although with reluctance. Coal was polluting, and it was costlier to obtain and distribute than wood. Coal was not available everywhere, and so entirely new distribution systems had to be devised. Digging a fuel from the ground costs more than cutting a standing tree, and coal overall costs more per unit of heating value than wood. People forced to heat with coal had to build chimneys in their houses, to vent the noxious smoke. Many of those forced to rely on coal thus experienced a decline in their financial well-being.

As coal gained importance in the economy the most accessible deposits were depleted. Mines had to be sunk ever deeper, until groundwater limited further penetration. This vexatious problem stimulated greatly the development of the steam engine, which in time was perfected enough to pump water from mines effectively. Thereafter the coal-based economy could not be turned back.

The fortuitous part of this transformation was that, with the development of an economy based on coal, a distribution system (canals and railways), and the steam engine, several of the most important technical elements of the Industrial Revolution were in place. It is one of history's great ironies that industrialism, that great generator of economic well-being, arose in part from steps taken to alleviate resource depletion, which is thought to produce poverty and collapse. The sense of irony is enhanced by the realization that both the wealth and the complexity of an industrial economy are made possible by the subsidy of a fossil fuel that people preferred not to use.

For a time, coal, steam engines, canals, railways, and industrial production methods interacted synergistically to produce increasing returns to complexity (the left side of Fig. 6.1 before point BI, CI). But those increasing returns were made possible by energy, and could not have occurred otherwise. We have not eliminated the problem of financing complexity. We have merely found energy slaves to bear the cost.

With this background we can now explore how energy and complexity have coevolved in history. Although we like to think of ourselves as unique, in fact our societies today are subject to many of the same forces and problems that past societies experienced, including problems of complexity and energy. In some past societies, the growth of complexity ultimately proved disastrous, and all past societies found it a challenge. Some found ways to pay for complexity, and at least one society abandoned it under necessity. Not only are we are not immune to the problems that affected past societies, we actually experience some of them in an even more intense form than did our ancestors. Because we have the gift of hindsight, it is important that we learn from these experiences. Therefore we turn now to discussions of energy and complexity in several past societies, returning to our situation today once some important lessons have been presented.

Collapse of the Western Roman Empire

The Roman Empire is paradoxically one of history's great successes and one of its great failures. The fact that it could be both of these things is due to changes in its complexity, the net returns to complexity, and its net energy. Early in Rome's history these factors were favorable and promoted imperial growth. Later they became unfavorable and made Rome vulnerable to collapse.

The story of the rise of the Roman Empire need not detain us long, interesting as it may be. The main lessons for today come later in its history. There are, however, a few points that it is important to understand.

Rome began as a small city-state on the banks of the Tiber River, controlling just a small amount of farmland in the vicinity of the settlement. According to Livy's history, these early Romans were perpetually at war with their neighbors, and many times their independence was threatened. In the end, though, they succeeded in overcoming these immediate threats. One of the persistent factors of Roman history is that success always brought new challenges. Conquering one neighbor always meant that the Romans now had a new neighbor, perhaps farther away but still threatening. Wars would beget more wars. There was never an end to the challenges. The only way the Romans could end a challenge was by conquering the challenger, and this they continually did. They were quite good at making war. In the centuries down to the third century B.C., after the Romans had conquered their immediate neighbors, they proceeded to conquer others who were farther away. In time they conquered or otherwise dominated most of present-day Italy.

Part of the Romans' secret was that, as they conquered more and more enemies, they would turn those enemies into allies. These allies were always subordinate to Rome, and were required to assist Rome in its ventures. The allies, in turn, survived being conquered and got to share in the spoils of further conquest. By the time that Rome engaged in wars of survival against

Carthage in the third century B.C., most of Italy and Sicily were united under its leadership. Rome by this time was a powerful state with very large reserves of manpower.

After Carthage was defeated in the Second Punic War (ended 201 B.C.), Rome was the most powerful state in the Mediterranean basin. Soon she was at war with Macedonia and Syria, successfully in both cases. Now Rome's wars started to become truly profitable. Increasingly the conquered peoples underwrote the costs of further expansion. These were societies powered entirely by subsistence agriculture, that is, by solar energy. There was not much ancient societies could do to store extra solar energy except to turn it into something durable. This they did by turning surplus solar energy into precious metals, works of art, and people. When the Romans conquered a new people, they would seize this stored solar energy by carrying off the same precious metals and works of art, as well as people who would be enslaved. Centuries of solar energy that had fallen on Mediterranean lands were seized and transported to Italy, making Rome the most magnificent city of the ancient world.

It is worthwhile to pause in our narrative and emphasize this point. The Romans' strategy of growth was to capture and use stores of past solar energy, stores that they did not have to create themselves. This is the same as we do today with fossil fuels. Nature has stored the past solar energy for us, whereas for the Romans it had been stored by the peoples they conquered. Both we and the Romans financed our growth with a subsidy that we did not have to produce ourselves. This analogy has lessons for our future that we discuss below.

We know some of the sums involved in Rome's conquests, and they seem staggering. The Romans minted a silver coin called a denarius. It was initially a coin of very pure silver, about 98–99%, having a diameter of about 18 mm and weighing a little under 4 grams. When the Romans annexed the kingdom of Pergamon in Asia Minor, they were able to double the state budget, from 25 to 50 million denarii per year. The Roman general Pompey raised the budget to 85 million denarii when he conquered Syria in 63 B.C. When Julius Caesar conquered Gaul (58–51 B.C.) he obtained so much gold that its value in Rome fell by 36%. With these windfalls, the Romans soon eliminated taxation of themselves.

One of the problems of being an empire is that eventually you run out of profitable conquests. Expand far enough and you will encounter people who are too poor to be worth conquering, or who are powerful enough that they are too costly to conquer. Diminishing returns set in. Rome reached its limits

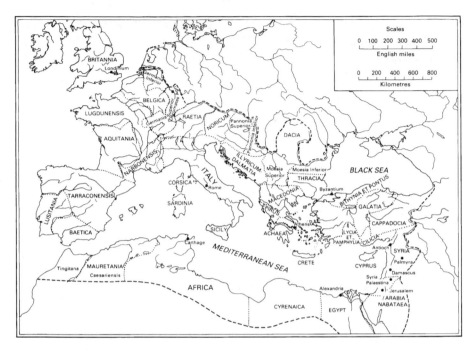

Fig. 6.4 The Roman empire in 117 A.D.

in northwestern Europe, where the peoples of Germany and Scotland were too troublesome and poor to be worth conquering, and in the east, where the Parthian Empire (of modern Iran and Iraq) was too powerful. Although there were some later conquests, Augustus, the first emperor (27 B.C.–14 A.D.), largely capped the size of the empire (Fig. 6.4). By the middle of his reign the state budget had increased to 125 million denarii.

The successes and failures of conquest make for fascinating history, but we are mainly concerned here with the economics of the Roman Empire. Rome was an agrarian society, in which farming made up 90% of the economy. It also made up 90% of the government's revenues. Trade and commerce were comparatively small parts of the economy. Most people were too poor to buy manufactured goods except for basics such as utensils for cooking and eating, and transport by land was too expensive for industry to thrive.

Thanks to various films, our image of the Roman Empire is that it was immensely wealthy and powerful. In fact it was more like a third-world economy of today. Only a few people were wealthy and powerful, and only some major cities were opulent. Everyone else lived a hand-to-mouth existence, and many people were in danger of hunger. When crops failed cities experienced famine. Unless a city was on the coast or a river, it was too expensive to transport grain to relieve a famine. Because of the threat of crop failures, farmers who owned their own land were always in danger of losing it.

The end of conquest meant that Rome's budget could no longer be financed from stored solar energy, by looting newly conquered peoples. Now the budget had to be financed from yearly solar energy, that is, from taxes on agriculture. During Augustus's reign, when funds sometimes fell short, he would relieve the state budget from his own wealth. That, however, came from the conquest of Egypt, another one-time infusion of stored solar energy. Later emperors, lacking Augustus's wealth, had to deal with increasingly intractable fiscal problems.

Roman silver and gold coins are found to this day as far away as the south of India, where they were traded for spices. The Romans also traded for silk from China. Merchants from foreign lands readily accepted these coins in the early empire because they could be trusted to be of pure metal and consistent weight. A king in India once expressed his admiration of Roman coins for these qualities. But good quality money comes from governments that are solvent, and solvent governments are one of history's rarest species. Funny money soon came to the Roman Empire. A problem faced by the Romans, as well as every other government, was how to meet fixed or increasing costs on the basis of fluctuating income. The empire's income fluctuated because nonmechanized agriculture always has variable yields. This meant that the Roman government rarely had reserves of cash to meet emergencies. By the early 60s A.D. Rome had been at war with Parthia for several years. Then in 64, Rome itself experienced the Great Fire, when Nero supposedly fiddled while Rome burned. These were expenses that could not be met out of ordinary income. Instead Nero resorted to a stratagem that later emperors found irresistible. He debased the silver currency, lowering the silver content from 98% to 93%. This was the start of a slippery slope (Fig. 6.5), and it resulted two centuries later in a currency that was worthless and a government that was insolvent.

Most of the history of the first two centuries A.D. need not concern us further. The empire's great crisis came in the half-century between 235 and 284. During this time the empire nearly came to an end. There were foreign and civil wars, which followed one upon another so often that there was almost no respite. Germans broke in from the north and Parthians from the east, devastating the provinces that they invaded. When emperors were not repulsing invaders they were fighting usurpers. It was a period of violent political instability. In this 50-year period there were at least 26 legitimate

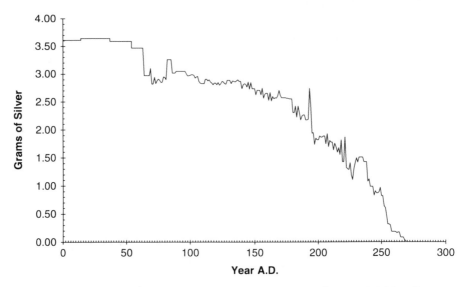

Fig. 6.5 Debasement of the Roman denarius to 269 A.D. (Source: T.F.H. Allen, J.A. Tainter, and T.W. Hoekstra. 2003. *Supply-Side Sustainability*. Columbia University Press, New York, p. 105)

emperors; they ruled an average of only 30 months. Histories of the period are so inadequate that the number of usurpers may never be known. Estimates run as high as 50, or an average of 1 per year.

In response to the crises, the army and bureaucracy grew in size and, because the government rarely borrowed money, taxes were increased. The Roman scholar Harold Mattingly wrote of this time, "The expenses of government were steadily increasing out of proportion to any increase in receipts and the State was moving steadily in the direction of bankruptcy." The silver denarius was replaced by a new coin, called today the antoninianus. The government issued it as worth two denarii, but it weighed only 1.5. Often the mint would take in denarii as taxes and immediately restrike them as antoniniani (Fig. 6.6), thereby doubling the government's spending capacity merely by issuing new coins. By 269 A.D. the once admired silver currency of Rome held only 1.9% silver, and this was primarily a silver wash that soon wore off in circulation. "The Empire," wrote Mattingly of this period, "had, in all but words, declared itself bankrupt and thrown the burden of its insolvency on its citizens."

To give examples of the many problems of this era: from the mid-240s until 272 there were continuous Germanic incursions, some reaching deep into Italy itself. In 247 the celebrations for Rome's 1,000th anniversary had to be postponed because the Emperor Philip was fighting in the Balkans. In 251 the emperors Trajan Decius and Herennius Etruscus were killed in battle with the Goths, along with much of their armies. Gallienus campaigned

Fig. 6.6 Fiscal distress in the Roman empire of the third century A.D. *Left*: denarius of Maximinus (235–238), minted 235. *Right*: antoninianus of Herennia Etruscilla (249–251), struck ca. 251 A.D. over a denarius of Maximinus. Maximinus' ear, flattened and inverted, can be seen to the right of the empress's ear, and under her neck are the letters XIM from Maximinus' name. The antoninianus was supposedly worth two denarii, yet the mint was taking denarii minted 16 years earlier and reissuing them as antoniniani. Maximum diameter of larger coin: 22 mm

yearly from 254 to 259 along the Rhine and the Danube, crushing a massive incursion of a tribe called the Alemanni in 259 at the very outskirts of Milan. His father, Valerian, was less fortunate. In 253 the Persians captured and sacked the eastern city of Antioch. Valerian went east, where he spent his reign campaigning. While he was occupied with the Persians, Goths attacked the undefended cities of Asia Minor. Finally in 260 Valerian was captured by the Persian King Shapur, and taken into captivity. He never returned.

It was the low point of Rome's fortunes, and for a time the empire disintegrated. The Roman Empire shrank to Italy, the Balkans, and North Africa. Even in this reduced empire there was much to do. In 267 the Goths returned to western Asia Minor and the Aegean, capturing and sacking Athens. In 270 an invasion of Germanic peoples again burst into Italy. Rome itself was defenseless, having long outgrown its ancient walls. The Emperor Aurelian (270–275) drove the invaders from Italy, but took care to see that the threat to Rome did not recur. He ordered that new walls be built around the city, the walls that are still seen today (Fig. 6.7).

Just when it looked like the Roman Empire might fall, the situation was rescued by a series of reforming emperors, most especially Diocletian (284–305) and Constantine (306–337). Their solution was to increase the size and complexity of the main problem-solving institutions, the government and the army. These emperors designed a government that was larger, more complex, and more highly organized. They commanded larger and more

Fig. 6.7 The walls of Rome

powerful armies. The government taxed its citizens more heavily, conscripted their labor, regulated their lives, and dictated their occupations. The most pressing need was a larger military. In 235 A.D. the army was about 300,000–350,000 men. It was doubled, to as high as 650,000, by the end of the fourth century. A second transformation was in the administration of the empire. Diocletian subdivided provinces into many smaller ones, and separated civil from military authority in each. This made it more difficult for provincial governors to rebel, but it also meant that there were many more provincial administrations. He increased the size of the imperial administration, which now moved with an emperor as he rushed to trouble spots. The bureaucracy was perhaps doubled in size.

The changes of Diocletian and Constantine made the empire more secure and efficient, but at substantial cost. Diocletian, in Edward Luttwak's words, "… turned the entire empire into a regimented logistic base …." He implemented Rome's first budget. The tax rate was established each year from a master list of the empire's resources, broken down province by province, city by city, field by field, household by household. Never before had the state so thoroughly penetrated its citizens' lives. Taxes apparently doubled between 324 and 364. Villages were held liable for the taxes on their members, and one village could even be held liable for another.

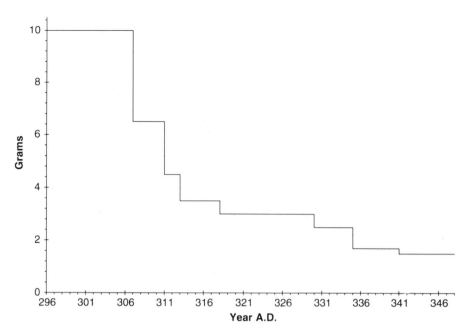

Fig. 6.8 Reduction in the weight of the Roman follis, 296–348 A.D. (Data from David Van Meter, 1991. *The Handbook of Roman Imperial Coins*. Laurion Numismatics, Nashua, NH)

Diocletian introduced a large silver-covered, copper coin, called today a follis. It started out weighing about 10 grams, with 4% silver, but it quickly began to shrink in size (Fig. 6.8). Roman emperors were often brilliant politicians and generals, but they were naïve economists. They did not understand that the amount of money in circulation affected prices. The continual debasements were inflationary, and the figures that have survived read like 1920s Germany. In the second century, a *modius* of wheat (about 9 liters) had sold during normal harvests for about 1/2 denarius. In 301 the price was set at 100 denarii. In 335 a *modius* of wheat sold in Egypt for over 6,000 denarii, and in 338 for over 10,000 (Fig. 6.9). In Egypt, an *artaba* of wheat (about 40 liters) had risen 27 times over the second century level, and from 250 to 293 the cost of a camel or donkey rose 60 times. In 301 a pound of pork was set at 12 denarii. By 419 it would cost 90 denarii. In the 150 years prior to Diocletian the value of gold had risen 45 times, the value of silver 86 times. Gold went from 50,000 denarii to the Roman pound in 301 to 504,000 in 450 (Fig. 6.10).

The strategy of the Roman Empire, in confronting a serious crisis, was largely predictable. The Romans responded as people commonly do: they increased complexity to solve their problems, and subsequently went looking

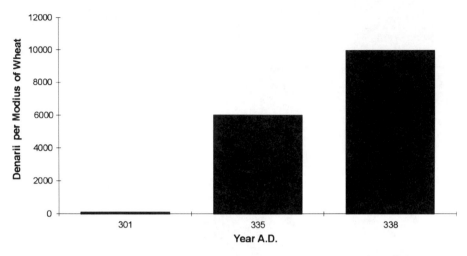

Fig. 6.9 Inflation in the price of wheat in Egypt. A modius was about 9 liters. (Data from: A.H.M. Jones, 1964. *The Later Roman Empire, 284–602: A Social, Economic and Administrative Survey.* University of Oklahoma Press, Norman, p. 119)

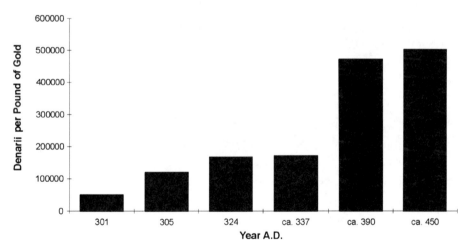

Fig. 6.10 Roman inflation: denarii to a pound of gold, 301 to ca. 450 A.D. (Data from: G.A.J. Hodgett, 1972. *A Social and Economic History of Medieval Europe.* Methuen, London, p. 38)

for the energy to pay for it. But in adopting this course the Romans found a dilemma that we will encounter in the future. Their energy budget was flat. It depended on solar energy, which could not be increased. This meant that to spend more on the army and the government, the Romans had to take resources from the peasants and, at the same time, further weaken their own finances. Both strategies deferred until the future the cost of current crises, rather as governments today routinely do. Will we eventually face a fatal problem such as the Romans did? Except for a brief bump in 2008, world oil production has been flat since 2005. We do not yet know whether this signals the peak of production that has been predicted, but that peak will come. In time the same will happen to coal and natural gas. As we discuss in Chap. 9, we may someday face the challenge of how to address expensive problems on a flat energy budget, just as the Romans did. This will have consequences for our ability to solve problems, for our solvency, and for the challenges that we leave for our children and grandchildren.

Although the extra army and government were needed, they were only the immediate tools of survival. Ancient states relied most of all on the peasantry, who supplied food for the cities, taxes for the state, and sons for the army. Here there were clear signs of trouble. In the late empire much cultivated land was abandoned. The tax system of the late empire seems to have been to blame, for the rates were so high that peasant proprietors could accumulate no reserves. Whatever crops were brought in had to be sold for taxes, even if it meant starvation for the farmer and his family. Farmers who could not pay their taxes were jailed, sold their children into slavery, or abandoned their homes and fields. The state always had a back-up on taxes owed to it, extending obligations to widows or orphans, even to dowries. The empire had suffered devastating plagues in the second and third centuries A.D., when large parts of the population died. Thanks to taxes, the peasant population failed to recover. Under such circumstances it became unprofitable to cultivate marginal land, as too often it would not yield enough for taxes and a surplus. Faced with taxes, a small farmer might abandon his land to work for a wealthy neighbor, who would be glad to have the extra labor.

As the economic basis of the empire deteriorated, it became increasingly likely that an insurmountable calamity would someday occur. The final downward slide began in 378, when an invasion of Goths destroyed the field army of the eastern empire. It was the start of a series of invasions that culminated in the fifth century with the establishment of Germanic kingdoms on what had been the territory of the western empire: Franks and Burgundians in the north of Gaul, Visigoths in the south of Gaul and in Spain, and Vandals in Africa. The Western Empire was by this point in a downward spiral. Lost or devastated provinces meant lower government income and less military strength. Lower military strength in turn meant that more areas would be lost or ravaged. The empire shrank eventually to Italy and adjacent lands. The Roman Army proper dwindled to nothing, and the army came to be staffed by German mercenaries. When those in Italy could no longer be paid, they overthrew the last emperor in the west in 476 A.D.

What the Roman Collapse Means for Us

The Roman Empire provides valuable lessons about energy and complexity, and about our possible future. We discuss two of those lessons here.

The rise and fall of the Roman Empire brings us back to the fungus-farming ants introduced in Chap. 5. Both ants and Romans, surprisingly enough, followed the same evolutionary course in energy and complexity. This tells us that we are dealing with universal constraints that affect all living systems, including us. We discuss here the results of a comparative study done by our colleagues Timothy Allen (University of Wisconsin) and Thomas Hoekstra (retired from the U.S. Forest Service) in collaboration with Tainter.

Early in their respective histories, both Romans and ants followed strategies of energy consumption that gave a high ratio of benefits to costs. Allen, Hoekstra, and Tainter have called this a high-gain strategy. In the case of the early Roman Empire, high gain came from employing the stored solar energy of conquered peoples to fund further expansion. In the last two centuries B.C., Roman expansion may have been nearly costless in an economic sense. Roman expansion was subsidized by centuries of stored solar energy.

It is likely that the earliest fungus-farming ants also employed a high-gain strategy to support their simple societies. We know from mitochondrial DNA that the earliest fungus-farming ants were like the species today that collect excrement, among other substances, as the substrate on which to grow their fungus. For ants, excrement is like jet fuel. Its content of nitrogen (the critical element) runs as high as five times that of leaves. By collecting substrate no more than 1 meter away from the nest, excrement collectors enjoy a benefit–cost ratio far greater than the leaf-cutters who haul small fragments of leaves up to 100 meters in long organized queues.

In human societies, at least, high-gain phases do not last very long. The resource seems inexhaustible, and so people expand their consumption of it. Yet the resource is finite, as the Romans found, and at some point expanding consumption crashes into a finite supply. Consumption must then shift onto lower-quality resources, initiating what we call a low-gain phase. There is a lower ratio of benefits to costs. The Romans shifted into low gain when they had to transition from using stored solar energy from conquests to consuming yearly solar energy from agricultural taxes. Fungus-farming ants transitioned into low-gain subsistence when new species emerged that used resources with lower nitrogen content, such as leaves.

Low-gain resources tend to be ubiquitous but widely dispersed, traits that characterize both peasant agriculture and leaves. Low-gain resources give small margins of net production, the excess of benefits over costs.

Peasant agriculture in the Roman Empire, for example, could produce about 1/2 metric ton per hectare and a yield of only three to four times the amount of seed planted. To accomplish much work at the societal level in a low-gain system, the small amounts of net production have to be aggregated. This requires high levels of organization. In agrarian empires, the organization comes in the form of elaborate systems to assess and collect taxes, and in bureaucracies to establish and administer tax systems. Among fungus-farming ants, low-gain organization takes the form of long lines of ants carrying leaves through tropical jungles, and the differentiation of leaf-cutter societies into distinct and highly efficient roles.

Since the emergence of coal as a steady resource we have been in a high-gain phase. Like the early Romans, we have been subsidized by stored solar energy in the form of fossil fuels. In typical fashion, we perceived fossil fuels as infinite, and expanded our societies to consume them. As we begin to fore-see the end of the high-gain fossil fuel era, many analysts feel that our future will depend on low-gain renewables. If it does, we will, like the Romans and the fungus-farming ants, transition to low-gain energy sources that yield small margins of net production. Our societies will rely on our own version of peas-ant agriculture and leaves. We discuss the implications of this in Chap. 9.

Another lesson from the Roman Empire is to understand one way of dealing with increasing complexity. We call it the Roman model. The society, in this model, increases in complexity to solve urgent problems, becoming at the same time increasingly costly. In time there are diminishing returns to problem solving, but the problems of course do not go away. The society must fund problem solving by extracting higher and higher amounts of resources, per-haps in the process degrading the productive system (the environment or taxpayers). In the Roman Empire this meant taxes on the peasants. But consuming more resources in the face of diminishing returns means that problem solving brings fiscal weakness, popular discontent, and ineffectiveness. If the society is lucky, the only problem will be ineptitude at solving problems. If it is unlucky, it will in time collapse, perhaps initiating a dark age. We show in Chap. 9 how complexification like the Romans practiced has led in our time to incidents such as the Gulf spill and our energy dilemma.

Collapse and Recovery of the Byzantine Empire

The debacle in Western Europe during the fifth century meant the end of the Western Roman state, but the Eastern Roman Empire (known today as the Byzantine Empire) persisted under its own emperors, changing greatly and coming to an end only when the Turks took Constantinople in 1453. For much of its history it lost territory, so that by the end the state consisted only of the city itself. Yet during the tenth and early eleventh centuries Byzantium was on the offensive, and doubled the territory under its control. There is a lesson in energy, complexity, and problem solving in the steps that made this possible.

Byzantine history has been a minor field of study in Western Europe and North America. Part of the reason is that the Byzantines were often on the defensive, and most people prefer to study and read about conquerors. Another is that the Western Roman Empire and its successors in Western Europe seem closer ancestrally than does Byzantium to western Europe and North America of today. This is a pity, for the history of Byzantium has drama that rivals that of any other place. To give just one example: in 685 A.D., a new emperor, Justinian II, ascended to the throne. His rule was oppressive, and he was overthrown in 695. To prevent him from reclaiming the throne, his nose was cut off, his tongue was cut (perhaps symbolically, inasmuch as he remained talkative for the rest of his life), and he was exiled to the Crimea. This would be enough to deter most men, but Justinian was made of solid stuff. The Crimean authorities soon tired of him, and planned to return him to Constantinople. They underestimated Justinian, who escaped to the steppes of Russia. There, even without a nose, he married the sister of a local khan. Justinian strangled with his own hands two assassins sent to kill him. Rallying some supporters, he sailed to the western Black Sea where he obtained an army from the emperor of the Bulgars. With a few followers he entered Constantinople in 705 through an unused aqueduct and seized the city. Despite his mutilation, he ruled until 711, when the Byzantines again rebelled against him, and this time put him to death. You cannot write fiction that is better than Byzantine history. As John Julius Norwich has said of the Byzantine emperors, "They were never, never dull."

Although it is tempting to indulge in stories about the escapades of Byzantine emperors, we must here, as with the Roman Empire, restrict our discussion to the prosaic topics of complexity and energy. Here Byzantine history truly shines. The Byzantines, as did the Romans, have lessons for us in these fundamental matters, lessons that tell us something about our future. The Byzantine lessons are singular, for Byzantium went through a transformation that may well be unique in the history of complex societies.

Following a series of economic and military reforms, the emperor Justinian (527–565) set out on an ambitious venture to reconquer the lost provinces of the West. An army sent to North Africa in 532 conquered the

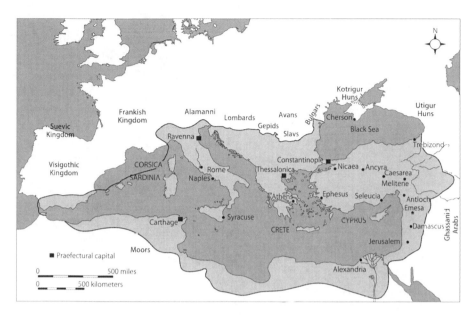

Fig. 6.11 The Byzantine empire in 565 A.D.

Kingdom of the Vandals within a year. Almost immediately, the Byzantine general, Belisarius, was sent to reconquer Italy. He had taken Rome and Ravenna, captured the Ostrogothic King, and conquered all of Italy south of the Po when he was recalled in 540 to fight the Persians (Fig. 6.11).

Then in 541 bubonic plague swept over the Mediterranean for the first time. Just as in the fourteenth century, the plague of the sixth century killed from one fourth to one third of the population. The loss of taxpayers caused immediate financial and military problems. The Lombards invaded Italy. In the early seventh century the Slavs and Avars overran the Balkans. The Persians conquered Syria, Palestine, and Egypt. Constantinople was besieged for 7 years.

This was a crisis in which the very existence of the empire was threatened. Responding to the military crisis and to reduced revenues, the emperor Heraclius cut pay by half in 616, and proceeded to debase the currency (Figs. 6.12 and 6.13). These economic measures facilitated his military strategy. In 626 the siege of Constantinople was broken. The Byzantines destroyed the Persian army and occupied the Persian king's favorite residence. The Persians had no choice but to surrender all the territory they had seized. The Persian war lasted 26 years, and resulted only in restoration of the status quo of a generation earlier.

The empire was exhausted by the struggle. Arab forces, newly converted to Islam, defeated the Byzantine army decisively in 636. Syria, Palestine, and Egypt,

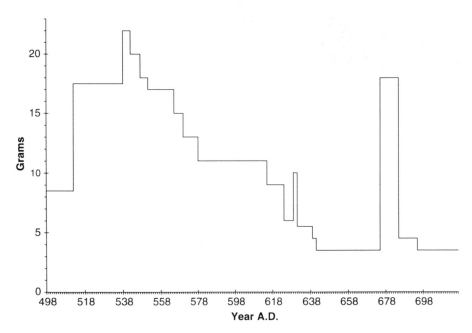

Fig. 6.12 Weight of the Byzantine follis, A.D. 498–717. (Data from: K.W. Harl, 1996. *Coinage in the Roman Economy, 300 B.C. to A.D. 700.* Johns Hopkins University Press, Baltimore, p. 197)

Fig. 6.13 Reduction of the Byzantine follis. *Left*: follis of Justinian, minted 538–539 A.D. *Right*: follis of Constans II, minted 655–656 A.D. Maximum diameter of larger coin: 40 mm

the wealthiest provinces, were lost permanently. The Arabs raided Asia Minor nearly every year for two centuries, forcing thousands to hide in underground cities (Fig. 6.14). Constantinople was besieged each year from 674 to 678.

Fig. 6.14 Rooms in an underground Byzantine city, Cappadocia, Asia Minor

Fig. 6.15 The land walls of Constantinople. (Source: Wikimedia)

The Bulgars broke into the empire from the north. The Arabs took Carthage in 697. From 717 to 718 an Arab force besieged Constantinople continuously for over a year (Fig. 6.15). It seemed that the empire could not survive. The city was saved in the summer of 718, when the Byzantines ambushed Arab

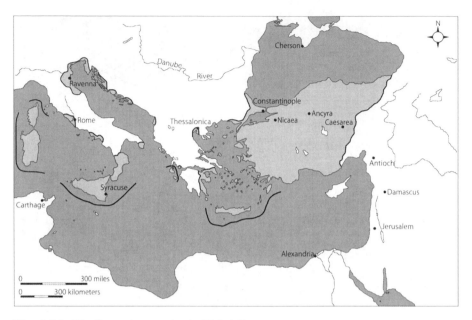

Fig. 6.16 The Byzantine empire in 700 A.D.

reinforcements sent through Asia Minor, but the empire was now merely a shadow of its former size (Fig. 6.16).

Third and fourth century emperors had managed a similar crisis by increasing the complexity of administration, the regimentation of the population, and the size of the army. This was paid for by such levels of taxation that lands were abandoned and peasants could not replenish the population. Byzantine emperors could hardly impose more of the same exploitation on the depleted population of the shrunken empire. Instead they adopted a strategy that is truly rare in the history of complex societies: systematic simplification.

Around 659 military pay was cut in half again. The government had lost so much revenue that even at one fourth the previous rate it could not pay its troops. The solution was for the army to support itself. Soldiers were given grants of land on condition of hereditary military service. The Byzantine fiscal administration was correspondingly simplified.

The transformation ramified throughout Byzantine society. Both central and provincial governments were simplified, and the costs of government were reduced. Provincial civil administration was merged into the military. Cities across Anatolia contracted to fortified hilltops (Fig. 6.17). The economy developed into its medieval form, organized around self-sufficient manors. There was little education beyond basic literacy and numeracy, and

Fig. 6.17 Gate to the Byzantine citadel of Ankara

literature itself consisted of little more than lives of saints. The period is sometimes called the Byzantine Dark Age.

The simplification rejuvenated Byzantium. The peasant-soldiers became producers rather than consumers of the empire's wealth. By lowering the cost of military defense the Byzantines secured a better return on their most important investment. Fighting as they were for their own lands and families, soldiers performed better.

During the eighth century the empire re-established control of Greece and the southern Balkans. In the tenth century the Byzantines reconquered parts of coastal Syria. Overall after 840 the size of the empire was nearly doubled. The process culminated in the early eleventh century, when Basil II conquered the Bulgars and extended the empire's boundaries again to the Danube. The Byzantines went from near disintegration to being the premier power in Europe and the Near East, an accomplishment won by decreasing the complexity and costliness of problem solving.

What the Byzantine Recovery Means for Us

Solar radiation reaches earth's upper atmosphere at a rate of 1.94 calories per square centimeter per minute. Thirty-one percent of this is reflected or scattered, and 23% is absorbed in the troposphere or upper atmosphere. The remaining 46% (about 0.9 calories) of the original solar radiation reaches the ground, or near it. Then, 34% of this is reflected back by snow or clouds. Forty-two percent goes to heat land and water. Twenty-three percent drives the water

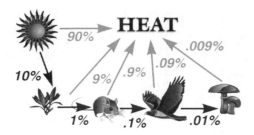

Fig. 6.18 Energy loss in ecosystems

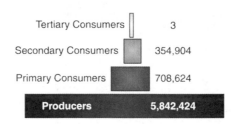

Fig. 6.19 A trophic pyramid

cycle, evaporation and precipitation. One percent drives wind and ocean currents. Of those original 1.94 calories, 0.023% is available for photosynthesis. That is 0.04 calories per square centimeter per minute to support nearly all life on earth, including everything that humans thought, did, and accomplished before we came to rely on fossil fuels. Of those 0.04 calories, the plant needs some for itself, so humans and other consumers actually get less. The wheel was invented, the pyramids were built, and Plato wrote philosophy on a small fraction of 0.04 calories per square centimeter per minute.

That is not the end of the story. Plants use solar energy, nutrients, and water to grow and reproduce. Herbivores eat the plants, carnivores eat the herbivores, and humans may eat all of them. As energy flows from plants to herbivores to carnivores, at each step about 90% of the energy is lost (Fig. 6.18). That is why top carnivores such as wolves and lions are rare. In an acre of bluegrass, for example, a population of six million plants can, in the end, support only three top-level carnivores (Fig. 6.19). The top of a food chain may run on 1/1000th of the energy that enters the ecosystem. In the top level of an ecosystem, individual organisms are rare but, to those who would consume them, highly nutritious. The nutrients have been concentrated and quality gained, even as quantity is lost. At the bottom of the food pyramid, numbers are vast but nutrient quality low. We see this in the difference between ants who collect insect droppings, which are high in nitrogen,

Energetics of the Byzantine Army

600 A.D.
Sun→crops→farmers→tax collectors→central government→army

700 A.D
Sun→crops→army

Fig. 6.20 Energy transformations in the Byzantine Empire, 600 and 700 A.D.

and ants that must transport endless quantities of low-quality leaf fragments. Leaves are ubiquitous, but it takes a lot of them to equal the nitrogen content of a single bit of excrement. In insect excrement, nature has concentrated the valuable substance. Similarly, fossil fuels are useful to us because geological processes have concentrated what we need from them, the capacity to generate heat.

This discussion of energy flow helps to clarify what the Byzantine Empire did to survive. The Byzantines responded to lower energy flow by simplifying their social, political, and economic systems, and by shortening energy flow networks. In the sixth century A.D., energy would typically have flowed from the sun to farms; from farms to peasants; from peasants to purchasers of grain, thereby converting the grain to coinage; then from peasants to tax officials; from tax officials to the government; and from the government to the army. At each step some energy was lost to transaction costs. Just as in an ecosystem's energy pyramid, energy was lost each time it was transformed or passed to another level (Fig. 6.20). After the mid-seventh century, the energy on which the empire depended flowed through fewer transactions and levels. Energy passed from the sun to farms. From there part of it was harvested and used directly by soldiers, eliminating many intermediate steps. Peasants still paid taxes, but because the government spent less of this revenue on the army there was more to allocate to other needs. Overall much less energy went to transaction costs, so net energy increased as a proportion of gross production. The empire revived and went on to expand.

The Byzantine recovery provides lessons for our future in two ways. First, a future in which we have less access to fossil fuels than we now enjoy will be a future dependent, at least in part, on renewable energy sources. These are mostly powered by the sun, either directly or indirectly. Yet the sun, as discussed above, can provide only 0.9 calories per square centimeter per minute at the ground, or 0.04 calories for photosynthesis. That is not a lot of energy, but if it is concentrated through taxes, in ancient times it could support an agrarian empire. But concentrating it involved bureaucratic salaries, transaction costs, and energy loss. The Byzantines found a way to minimize these costs. The Byzantines shifted the bulk of their military activities down the political food chain, transforming soldiers from being consumers to producers of the empire's wealth. Second, the empire was forced to simplify in a manner consistent with its available energy. We term this the Byzantine model: recovery through simplification. It is a solution that is often recommended for modern society as a way to inflict less damage on the earth and the climate, and to live within a lower energy budget. The Byzantine Empire is, to our knowledge, the only large complex society that has actually done this, that has survived by simplifying to live within the constraints of less available energy. In this sense, Byzantium may be a model or prototype for our own future, in broad parameters but not in specific details. There is both good news and bad news in this. The good news is that the Byzantines have shown us that a society can survive by simplifying. The bad news is that they accomplished it only when their backs were to the wall. They did not simplify voluntarily.

It is worthwhile to pause a minute and reflect on modern agiculture. On the year-around average, about 200 watts (or Joules per second, enough to power two people; see Chap. 3) of solar radiation impinges on each square meter of flat land in the United States. In a very good year, U.S. agriculture – the most efficient in the world – sequesters about 0.36 watts per square meter as grain and seeds, and 0.66 watts per square meter as all above ground biomass and roots. Thus, U.S. agriculture sequesters only three parts in 1,000 of all solar light falling year-round on the soil surface. Compared to what we are used to with fossil fuels, solar energy is not a very productive basis for a society.

Warfare and the Development of Modern Europe

Here we come to a historical case that leads directly to our own time. It also provides us with unique insights into how we have come to live as we do, including our use of energy.

Fig. 6.21 Tarascon castle, Provence, France, built in the thirteenth century

Arms races are the classic example of diminishing returns to complexity. Any competitive nation will quickly match an opponent's advances in arma ments, personnel, logistics, or intelligence, so that investments typically yield no lasting advantage or security. The costs of being a competitive state continuously rise, and the return on investment inexorably declines. Arms races provide a classic illustration of increasing in complexity and costliness in order to maintain the status quo, to remain secure. The development of Renaissance and modern warfare illustrates this process.

In Europe of the fifteenth century, siege guns ended the advantage of stone castles (Fig. 6.21). Fortifications were developed that could support defensive cannon and that could also survive bombardment. These new fortifications featured low thick walls with angled bastions and extensive outworks (Fig. 6.22). They were effective but expensive: Siena built such fortifications against Florence, but was annexed anyway when no money was left for its army.

Open-field warfare also developed greater complexity. Massed archers and the pike phalanx made the armored knight obsolete. These were soon superseded by firearms. Effective use of firearms took organization and drill. Victory came to depend not on simple force, but on the right combination of infantry, cavalry, firearms, cannon, and reserves.

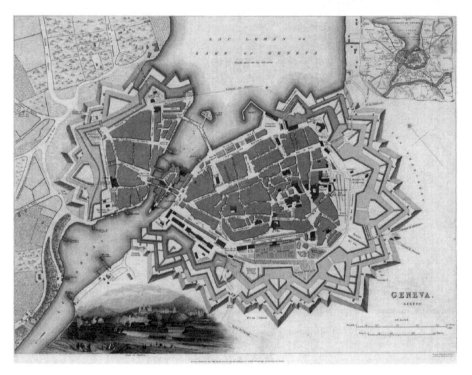

Fig. 6.22 Plan of Geneva in 1841, showing *trace italienne* fortifications. (Source: Wikimedia)

War came to involve ever-larger segments of society and became more burdensome. Several European states saw the sizes of their armies increase tenfold between 1,500 and 1,700. Yet land warfare became largely stalemated. The new technologies, and mercenaries, could be bought by any power with money. When a nation threatened to become dominant, alliances would form against it. Defeated nations quickly recovered and were soon ready to fight again.

Warfare evolved into global flanking operations. The development of sea power and acquisition of colonies became part of stalemated European warfare (Fig. 6.23). Yet expanding navies entailed further problems of complexity and cost. In 1511, for example, James IV of Scotland commissioned the building of the ship *Great Michael*. It took almost one-half of a year's income to build, and 10% of his annual budget for seamen's wages. Ships thereafter continued to grow in complexity and cost.

In 1499 Louis XII asked what was needed to ensure a successful campaign in Italy. He was told that three things alone were required: money, money, and still more money. (The reply might have been: energy, energy, and

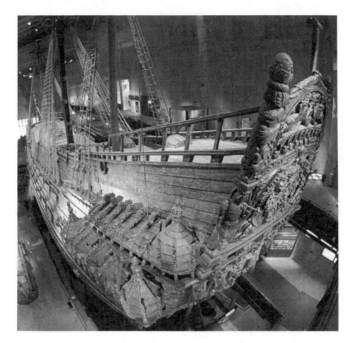

Fig. 6.23 The seventeenth century Swedish warship *Vasa*

still more energy). As military affairs grew in size and complexity finance became the main constraint. The cost of putting a soldier in the field increased by 500% in the decades before 1630. Nations spent more and more of their income on war, but it was never enough. The major states had to rely on credit. Even with riches from her New World colonies, Spain's debts rose 3,000% in the century after 1556. War loans grew from 18% interest in the 1520s to 49% in the 1550s.

The wars raised permanently the cost of being a competitive state, and war-induced debt levels persisted long after the fighting ceased. Power always shifts, and victorious nations were never able to dominate for very long. Many people of the time understood the futility of European wars, but arms races are especially difficult to break. In 1775 Frederick the Great eloquently described the state of affairs.

> The ambitious should consider above all that armaments and military discipline being much the same throughout Europe, and alliances as a rule producing an equality of force between belligerent parties, all that princes can expect from the greatest advantages at present is to acquire, by accumulation of successes, either some small city on the frontier, or some territory which will not pay interest on the expenses of the war, and whose population does not even approach the number of citizens who perished in the campaigns.

Europe's continual wars had to be supported by an economy based largely on solar energy, and taxes paid by a peasantry that grew ever more desperate. If there was ever a political system that should have been vulnerable to collapse from its own costs, it was Europe of the last millennium. Why did this not happen?

There are two primary reasons why today's prosperous Europe emerged from so many centuries of misery. The first is that the competition forced Europeans continuously to innovate in technological prowess, organizational abilities, and systems of finance. They were forced to become more adept at manipulating and distributing matter and energy. The second reason is that they got lucky: they stumbled upon great subsidies. Over the seas they found new lands that could be conquered, and their resources turned to European advantage. We are all familiar with the stories of untold riches that Europeans took from the New World. To this day, treasure hunters still search the Caribbean for Spanish ships. These riches were, of course, transformed solar energy.

European competition stimulated great complexity in the form of technological innovation, development of science, political transformation, and global expansion. To subsidize European competition it became necessary to secure the produce of foreign lands. The energy and resources of other peoples were channeled into this small part of the world. The maritime European powers, as did the Romans before them, seized the solar energy of vast territories and used it for themselves. This concentration of global resources allowed European conflict to reach heights of complexity and costliness that could never have been sustained with European resources alone.

What Warfare and the Development of Europe Mean for Us

Increasing complexity on a fixed national budget amounts, as the Romans showed us, to robbing Peter to pay Paul. As the portion of national energy that goes to one sector, the military in this case, increases, other sectors must get smaller shares. If not, then increasing complexity will actually produce diminishing capability, as arms races sometimes illustrate.

Competition spurs complexity, as each competitor seeks to outdo its rivals. For the most part we consider competition a good thing: competition brings us improved products at lower cost, and overall a better material quality of life. As we know by this point, complexity carried too far becomes unwieldy and too expensive. In a competitive system such as an arms race this produces what we call an "escalation dynamic." Military hardware grows increasingly complex and capable, but also increasingly costly. The growing

Fig. 6.24 A B-2 bomber being refueled. (Source: Wikimedia)

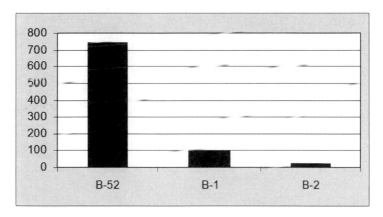

Fig. 6.25 Production of strategic bombers

costs mean that fewer and fewer weapons can be procured. The United States ended World War II, for example, with a fleet of 32 aircraft carriers, and with several more in production. Today the U.S. Navy struggles to support 11 carriers. These are, to be sure, highly capable weapons platforms, but 11 carriers cannot be in 32 places at once. In this sense, the cost of complexity constrains capability. Similarly, in the early years of the Cold War, the U.S. produced 744 B-52 strategic bombers. Many of them are still flying today, and some are expected still to be flying in the 2030s. When the B-1 bomber entered production in the 1980s, only 100 were made. Of the most recent strategic bomber, the B-2 (Fig. 6.24), merely 21 were produced (Fig. 6.25).

Those are 21 amazing planes, but they can only be in 21 places at the same time. The military refers to this process as the death spiral. Because of high costs, fewer planes are bought. The cost of each plane rises, even fewer planes are bought, and the unit cost rises still further. It is an inevitable outcome of complexification, whether in military technology or in other areas of our lives. Wherever we must escalate technology to accomplish our goals, even to maintain the status quo, the increasing costs will constrain capability. No matter how many deep-sea drilling rigs we might need, we cannot afford very many if they cost $1,000,000,000 apiece.

We call this process the European model of increasing complexity. Problem solving produces ever-increasing complexity and consumption of resources, regardless of the long-term cost. European warfare provides a valuable lesson in how to support complexity, lessons that extend beyond military competition and apply to supporting complex societies in general. High complexity in a way of life can be sustained if one can find a subsidy to pay the costs. This is what fossil fuels have done for us: they have provided a subsidy that allows us to support levels of complexity that otherwise we could not afford. In effect, we pay the cost of our lifestyle with an endowment from a wealthy ancestor. That ancestor is the geological stores of eons of past solar energy, transformed into petroleum, natural gas, and coal. This is fine, as long as the subsidy continues undiminished and as long as we do not mind damages such as the Gulf oil spill. We would do well, though, to keep in mind the experience of the Romans, who found that high-gain subsidies do not last.

Conclusions

It is important to emphasize a point made earlier in this chapter. Complexity is not inherently good or bad. It is, rather, effective and affordable, or it is not. One of the main ways that complexity increases is to solve problems. As long as the benefit–cost ratio is favorable, complexification is indeed an effective way to solve problems. That is why we adopt it so readily. It has served us well until now. Yet as emphasized in this chapter and the last one, complexity is not free. It always costs, and the ultimate cost is energy. As we become a more complex society, we consume more energy. Aside from the enormous amounts of energy that we must continually find and produce, complexity presents us with a second dilemma: the ratio of benefits to costs is not constant. Complexity as a problem-solving strategy reaches diminishing returns.

As the most cost-effective ways to solve problems are progressively exhausted, those remaining tend to cost more and give lower net benefits. The effectiveness of complexity declines.

We see in the historical case studies different options to the process of complexification. In the Roman model, problem solving drives increasing complexity and costs that cannot be subsidized by new sources of energy. In time there are diminishing returns to problem solving. Problem solving continues by extracting higher levels of resources from a fixed energy budget. It is a strategy of robbing Peter to pay Paul. Fiscal weakness and disaffection in time compromise problem solving and initiate collapse.

In the Byzantine model, the society simplifies deliberately and systematically when faced with insufficient energy. It is a strategy that can succeed, but it too has a cost. That cost is the need to give up an accustomed way of life. People will generally accept this only when it is clear that there is no alternative.

Finally there is the European model, in which we still participate. Problem solving drives ever-increasing complexity and consumption of resources. A society can follow this strategy over the long term only if it finds an abundant source of energy to subsidize the cost of complexity. This has been our strategy for the past 200 years. The question is how long this strategy can continue. For 200 years we have been living off the geological equivalent of an endowment from a long-dead ancestor. That endowment will not last forever, and in fact it is becoming harder and harder for us to access the amount of it that remains.

These models provide us with a framework for thinking about energy and our future. They allow us to see options and clarify alternative paths. We know that these outcomes can happen, because they have already happened. As we discuss in Chap. 9, our alternative futures could involve collapse, deliberate simplification, or further complexification based on a new energy subsidy.

Our infrastructure to find and produce oil has undergone an evolutionary course like that described in this chapter. Where we once got oil by sticking a straw in the ground, now it takes $1,000,000,000 contraptions such as the Deepwater Horizon. Our efforts to find and produce oil have experienced increasing complexity and costliness, and have produced diminishing returns. Diminishing returns in the petroleum industry also come from declining net energy. As we deplete the easiest reservoirs of oil, the remaining oil is more inaccessible and costlier to produce. The technology that we need to find and produce it grows more complex and costly. It also grows more risky to use. This means that it takes more energy to produce the amounts of energy that our societies require. Net energy declines. Theoretically, it could someday take a barrel of oil to get a barrel of oil. That will never happen, of course. We will be forced to stop using petroleum long before that day comes.

The last word on our civilization has not been written. The Deepwater Horizon disaster is only one episode along a collision course between the energy we require, the increasing risks incurred in finding and producing oil, and the net energy that we get in return. Was the well blowout and its aftermath an accident, or a nearly inevitable consequence of the energy–complexity spiral? Can our society reach into the bag of complexity tricks yet again to avoid future blowouts? Does our very approach to solving problems make major accidents all but inevitable? Will complexity have the same consequence for us that it had for past societies? We return to the blowout itself in the next two chapters, then take up complexity again in Chap. 9. Unfortunately, complexity at every level of offshore drilling, from the technology to the board room to onsite decision making, means that the scope for other potentially catastrophic errors is very great, perhaps even predictable.

Further Reading

Tipping Points

1. Gladwell, M.: The Tipping Point: How Little Things Can Make a Big Difference. Little Brown, Boston (2000)

Benefits and Costs of Complexity

2. Tainter, J.A.: The Collapse of Complex Societies. Cambridge University Press, Cambridge (1988)

Case Studies

3. Allen, T.F.H., Tainter, J.A., Hoekstra, T.W.: Supply-Side Sustainability. Columbia University Press, New York (2003)
4. Patzek, T.W.: How Can We Outlive Our Way of Life? OECD Paper, Paris, www.oecd. org/dataoecd/2/61/40225820.pdf. (2007). Accessed 6 August 2011
5. Smil, V.: Energy in World History. Westview, Boulder (1994)
6. Tainter, J.A., Allen, T.F.H., Little, A., Hoekstra, T.W.: Resource transitions and energy gain: contexts of organization. Conserv. Ecol. 7(3), 4 (2003). http://www.consecol. org/vol7/iss3/art4

Chapter 7

What Happened at the Macondo Well

April 20th was a calm sunny day in the Gulf of Mexico. The 126-strong crew of the Deepwater Horizon drilling rig was busy finishing work on the BP Macondo well at a remote location, as shown in Fig. 7.1, some 130 miles southeast of New Orleans and 420 miles east of Houston. Only a few routine operations were left to be completed before the rig would leave the site, with the Macondo well safely plugged with cement and waiting to be reopened at a later date. By March 8, the rig was already scheduled to be at a different location in the Gulf, so everybody was rushing about to mop up and stop the hemorrhage of money escaping from BP's pocket at a rate of $1,000,000 per day.

Hovering almost 1 mile above the seafloor, the rig was connected to the well through a 18-3/4 inch-wide umbilical cord, which the industry refers[1] to as the "riser pipe" or "drilling riser." The automatic positioning systems on the rig made sure that the riser was neither bent nor flexed too much. Through flexible joints and computer-controlled cranes, the top end of the huge riser was connected to the drilling floor on the rig. The riser's bottom end was attached to a special device, the "lower marine riser package," or just LMRP, whose two "annulars" controlled the flow up and down the riser. Inserted into the riser was an assembly of smaller 6-5/8, 5-1/2, and 3-1/2 inch

---

[1] Many of the industry-specific terms are explained in the Glossary at the end of this book.

J.A. Tainter and T.W. Patzek, *Drilling Down: The Gulf Oil Debacle and Our Energy Dilemma*, DOI 10.1007/978-1-4419-7677-2_7, © Springer Science+Business Media, LLC 2012

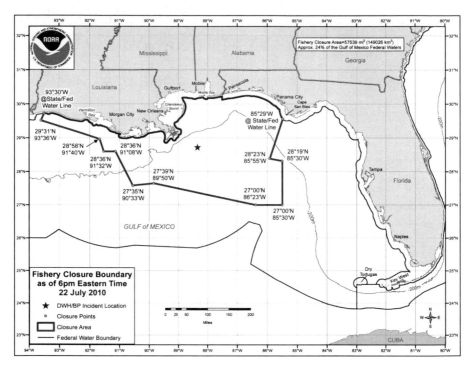

Fig. 7.1 The location of the BP Macondo well and some 50,000 square miles closed to all fishing on July 22, 2010, 3 months after the spill. The beaches in the closed-off area have been contaminated with oil. The degree of contamination ranged from heavy to no oil, and assumed a more or less random pattern. (Source: The NOAA Fisheries Service)

drill pipes[2] that extended down 3,300 feet into the well to transport drilling mud, cement, and seawater from the ship. These drill pipes were to be used later that night to inject a heavy mud, conduct a "negative pressure test," and secure the well with a second cement plug.

The LMRP also provided a flexible connection between the riser and the well that, in case of trouble, would allow a fast disconnect from a stack of powerful valves and shears screwed into the wellhead just above the seafloor. This "blowout preventer stack," or BOP, that was all but ubiquitous in news coverage,

---

[2] According to BP, the 6-5/8 inch pipe went down to 4,117 feet, 960 feet above the seafloor. The 6-5/8 inch pipe's bottom end was then joined with a 5-1/2 inch pipe down to 7,567 feet, followed by a 3-1/2 inch pipe down to 8,367 feet, or 3,300 feet/1 kilometer into the wellbore. The reason for the three different pipe diameters was that the two larger diameter pipes would get stuck in the central production casing of the well. According to Halliburton, there was no 5-1/2 inch pipe and the 6-5/8 inch pipe went down to 7,545 feet.

controlled flow in and out of the well. Together, the BOP and LMRP were the height of a five-story building, weighed some 450 tons, and represented an awesome feat of engineering, particularly considering that both devices had to be lowered precisely onto the 22-inch wellhead a mile below the Gulf surface.

At 5 p.m., the crew started the negative pressure test to check for leaks from the well through any of its pipes. At 5:15 p.m., 454 barrels[3] of the heavy and viscous "lost circulation mud" were pumped through the drill pipe in the riser into the wellbore and the riser's bottom-end. Nobody knew exactly where these 454 barrels of mud flowed and how they mixed with the ensuing seawater, making the negative pressure test confusing and difficult to interpret. From the pressure readings on the rig, the crew concluded, with tragic consequences, that the well was intact (Figs. 7.2 and 7.3).

We return to the results of the negative pressure test a little later, but first let's consider a few other aspects of this test that shine more light on what actually happened. With written permission from the Minerals Management Service (MMS), the drill pipe assembly protruded 3,300 feet into the well, much deeper than the customary 1,000 feet used for pressure testing. This pipe assembly was used to pump the heavy mud first, and then seawater into the large central tube running down the length of the well called the "production casing." The injected seawater displaced the heavy drilling mud which up to that moment had filled parts of the production casing and riser. The effect of the seawater was to make the production casing lighter and more buoyant. Also the hydrostatic fluid pressure at the bottom of the casing decreased at first[4] by about 1,300 psi, making it a whole lot easier for the reservoir fluid to break into the casing. If the drilling mud in the drill pipe, casing, and riser were replaced by seawater all the way to the ship, the pressure decrease at the bottom of the casing would be a whopping 2,360 psi.[5]

As described below, Halliburton's cement was likely unstable, sensitive to contamination by the oil-based drilling mud and reservoir fluid, might have degassed catastrophically, and likely did not set properly. The reservoir fluid might have created a path flow through the cement paste inside the casing, and then flowed through the two mechanical barriers (the "floats") that did not close because of yet another serious error. This latter error was most likely caused by a

---

[3] One barrel is 42 U.S. gallons or 159 liters.

[4] Displacing 16 pounds per gallon (ppg) mud with 8.6 ppg seawater.

[5] Displacing a 14 ppg mud with 8.6 ppg seawater. Because some of the riser was filled with a 16 ppg mud, the actual pressure decrease was even higher.

Fig. 7.2 You can think of the BP Macondo well as a very slim telescope, 13,300 feet long, made of eight segments of slender steel pipes of increasingly smaller diameters. The ninth, narrowest pipe, called "production casing" runs at the center, along the entire length of the telescope. This telescope points vertically upwards and is cemented in a borehole drilled with bits of different diameters. The wide end of the telescope is the 36 inch outer diameter (OD), 250 feet long conductor casing, driven into the sea bottom with water jets. The conductor casing is not cemented in place and its role is identical to that of the iron pipe in Colonel Drake's Titusville well. The top ends of the next two pipes are flush with the wide end of the telescope, are 1,150 and 2,900 feet long, and have outer diameters of 28 and 22 inches, respectively. These two pipes are cemented in place. A flange at the top of the 22-inch pipe becomes the "wellhead." The other five, progressively narrower, segments of casing, are hung off the wider ones, just like the segments of the telescope. Each of these hanging pipes is called a "liner," not "casing." The exceptional 16 inch liner is 6,360 feet long, and is hung just 160 feet below the well top. The narrowest and deepest pipe (the eye-piece end) is 7 inch OD and extends all the way to the top. The 7-inch casing widens to 9-7/8 inch OD at 5,870 feet above the bottom

decision by BP to limit the flow rate through the casing in order to prevent further rock fracture during cementation. This rate, four barrels of slurry per minute, was less than the minimum flow rate of six barrels per minute necessary to convert the floats, so that they could block fluid flow up the casing.

If the cement did not set inside the casing and bond properly to the casing wall, then it would not set in the borehole as well, allowing the reservoir fluid eventually to work its way up from the depth of 18,360–17,168 feet, where this

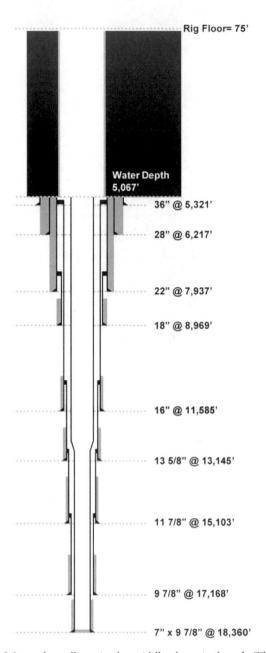

Fig. 7.3 The BP Macondo well cut in the middle along its length. The thick gray areas denote cement between the outer walls of the different pipe segments and the borehole. Note that the well is *not* cemented from top to bottom and there are large intervals where the steel pipe contacts rock in the borehole. After the conductor casing and the 28 and 22-inch casing pipes were in place, a blowout pressure preventer was installed, and the remaining drilling, casing, and cementing operations were conducted through a riser pipe connecting the BOP with the drillship. 7″ × 9 7/8″ means the 7 inch OD production casing tube becoming the 9-7/8 inch tube at the depth of 13,145 feet. See Fig. 7.2 for a simple explanation of well architecture. (Source: Halliburton)

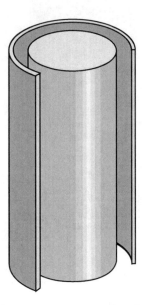

Fig. 7.4 The annular space between two concentric tubes. The inner tube is the production casing and the outer tube is the well casing cemented inside an even larger cylindrical borehole

fluid could breach the bottom seal of the well's annular space. However, this possibility now appears unlikely given the evidence from the Macondo well blowout preventer that was recovered from the seafloor. Either way, the reservoir fluid formed large bubbles that were able to flow by buoyancy all the way up to the wellhead and enter the blowout preventer. Once the reservoir fluid started flowing through the casing, plenty of gas would evolve from the decompressing light, gas-rich oil. This gas then rushed up the blowout preventer and the riser, arriving some 18 minutes later at the ship 1 mile above the wellhead. As it traveled up, the evolved gas expanded its volume more than 100 times and hit the ship's bottom with the force of a speeding train. At that moment, there was probably nothing one could do to save the ship.

In May 2010, shortly after the blowout, few knew about the problems with cement across the bottom 190 feet of the casing, and no one knew that the valves ("floats") that were expected to stop fluid flow up the casing were not activated. In addition, no one knew if the blowout preventer had closed or not. Therefore, most engineers focused on a breach of the liner seal at the depth of 17,168 feet, and the subsequent invasion of the "annular space" of the well, depicted in Fig. 7.4, by reservoir fluid. Because BP did not lock

Fig. 7.5 *Top:* The production casing hanger seal assembly from the outside. Apparently, the Macondo well casing did not move up and then drop down, because the seal is intact, as the two arrows indicate. Also, if the reservoir fluid flowed through the annular space, it would have to exit through the small holes on the perimeter of the assembly. The four visible holes are not eroded. *Bottom:* An image of the inside of a new casing hanger assembly (*left*) and in the Macondo well (*right*), showing significant erosion due to flow through the production casing. The *bottom arrow* shows the eroded square tab on the left. The *upper two arrows* highlight the extent of erosion of the grooves in the hanger. (Image source: Pages 118 and 119 of the presentation based on the *Deep Water – The Gulf Oil Disaster and the Future of Offshore Drilling*, Report to the President, National Commission on the BP Deepwater Horizon Oil Spill and Offshore Drilling)

down the production casing to the wellhead, it was surmised that a very high pressure of the reservoir fluid in the annular space pushed the casing up, causing the casing seal to be breached at the wellhead.[6]

From the published analyses of the cement slurry used by Halliburton to seal the casing, and the images of the production casing and casing hanger shown in Fig. 7.5, we learned that in fact the bottom of the casing was

---

[6]Take an empty tall metal can. Press it down with your hands until it partially submerges in water. Then let it go. The can will jump up from the water. So could the production casing in BP's well when pressure was decreased in the riser and both annular preventers were opened.

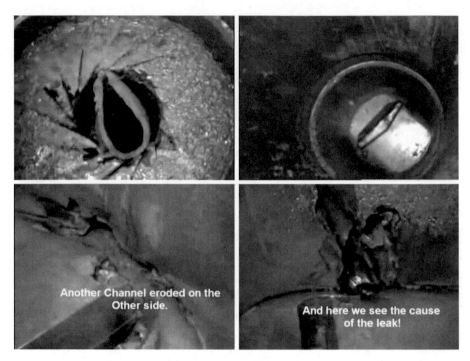

Another Channel eroded on the Other side.

And here we see the cause of the leak!

Fig. 7.6 Four frames from a video of an inspection of the Macondo well Blowout Preventer lifted up by the Helix Q4000 vessel on November 6, 2010. From the upper-left, clockwise, the fiber-optics video camera shows (1) the closed annular preventer with a drill pipe crimped in, (2) the almost-closed and eroded blind shear ram below, (3) erosion channels on one side of the ram, and (4) on the other side of it. Reservoir oil and gas flowed through a narrow slit in the ram and etched the hard steel shears. The video was leaked to YouTube, www.youtube.com/watch?v=UxCt3UsmJF0

breached, because the Halliburton cement was too weak and the valves that were supposed to stop the flow were open. We also know now (Det Norske Veritas report listed in the Further Readings section) that the blowout preventer stack almost closed, albeit likely with some delay (Fig. 7.6). The reservoir fluid that carried sand started flowing through the narrow slit in the almost-closed blind shear ram that did not quite cut through the off-center, bent drill pipe, and scoured deep grooves in the hard steel of the ram shears. The incredible video posted on YouTube lets you see this for yourself.

The BOP acted as an effective barrier to flow and a choke on the wide-open production casing. In the early days, the BOP was virtually closed and the flow rate out of the Macondo well was low. But with time, the downhole valves and the BOP's blind shear ram were eroded away by the reservoir fluid and sand, and the flow rate increased, perhaps several-fold. Below, we discuss the all-important question of a possible range of the flow rates out of the Macondo well.

Returning to the negative pressure test, the contents of the drill pipe, 3,330 feet of production casing, and the entire riser pipe were flushed out with seawater between 6:29 and 9:47 p.m., removing all of the heavy mud above the wellbore and in the upper quarter of the production casing. The well contents were now lighter and exerted insufficient pressure on the cement blocking the bottom 189 feet of the casing from an invasion by the overpressured reservoir fluid. The weak cement was crushed by the fluid that flowed up the casing to the wellhead. The annulars were open, and the oil and gas that filled the casing could flow all the way up to the rig. It would take only 18 more minutes for the gas to travel the mile from the wellhead to the rig's drilling ("derrick") floor, where a few crew members were busy pumping the contents of the riser and a mud tank overboard to another ship docked at the rig's side. No one was watching the mud level in the tank and no one was observing the increasing flow rate[7] from the well. Suddenly, there was a loud hissing sound, and the crewmen on the other ship saw the mud and water spraying out of the transfer pipe and raining down on their heads.

Below the rig's deck, senior members of the crew and a couple of BP VIP's, who had flown to the rig that day, were celebrating their impeccable safety record.

Two minutes later, at about 9:49 p.m., the hissing sound became a roar as loud as a jet plane taking off, and mud and seawater shot up some 300 feet above the derrick floor. The riser pipe exploded under the enormous pressure of gas and oil racing up to the rig from the breached well. In an instant, from sea level up to the main deck, the Deepwater Horizon was enveloped in a huge gas cloud with tiny water droplets shrouding it like a fog.

Moments later there was a huge explosion and a fire ball visible from tens of miles away. After several more explosions, the Deepwater Horizon became an inferno with 11 crewmen dead or dying, and 115 stumbling in total panic onto the rig's lifeboat deck or jumping into the sea.

Flames heated the air and illuminated debris rained back down onto the rig. Workers screamed and yelled, some trying to organize an evacuation, others simply unable to bear the pain of their wounds. Almost miraculously,

---

[7]The computerized sensors were registering the flow rate, but their readings were not interpreted correctly or fast enough to make a difference.

Fig. 7.7 The Transocean semisubmersible drilling rig is sinking almost exactly 2 days after the initial explosions. (Source: U.S. Coast Guard – 100421-G-XXXXL- Deepwater Horizon fire, displayed in Wikipedia)

115 people survived. Although the gripping pictures of smoke and fire tell us little about the cause of the disaster, they leave no doubt why it only took 2 days for the giant rig to sink to the bottom of the Gulf (Fig. 7.7).

The voices of three rig workers were caught on tape by reporters from *The New York Times*, and the recordings transcribed here were posted to www.nytimes.com/interactive/2010/05/-08/us/20100508RIG_CLIPS.html on May 8, 2010. Let's listen in to these first-hand accounts.

Mr. Eugene Dewayne Moss, Crane Operator on the Deepwater Horizon rig:

To tell you the truth, I did not have time to get scared. All I could think of was my wife and kids. If you ever heard a tornado before, that's sort of what it sounded like. That's about the time I felt the first explosion. The lights went out at the second explosion. On the second one I had to hold on to the wall. Then we knew that something was badly wrong. The first thing that shocked me was that the ceiling had collapsed onto the stairway. And this is when you knew this is the time to get out of town. I saw the whole deck on fire and this was pretty shocking. I saw people jumping from the deck. Actually it is truly horrifying because you cannot see them good enough if they had life jackets or anything like that. All you could see is figures jumping from the roof into the water. All I could do was pray.

Mr. Micah Joseph Sandell, Crane Operator on the Deepwater Horizon rig:

When I first saw mud coming out, I knew we was having a blowout. What went though my mind was: "That mud to clean up now and they'll get control of it." And when it quit that's when I took a deep breath thinking that they got the control of it. Then when gas came out, I've never seen that before. I've never seen that kind of pressure before. And what scared me, I didn't know if I should run or not. A minute or so after it was a big explosion. The whole deck blew up. Fire shot up all the way up to the top of the derrick, 300 feet in the air. Fire went around me and blew me to the back of the cab. When I hit the back of the cab, I just fell down and I put my arms over my head. To be honest with you, my only word I remember saying right then was, "No, God, no!" because I felt that was it for me because there was fire all around me.

Mr. Dennis Dewayne Martinez, Supervisor on the Deepwater Horizon rig:

[When the first blast came] I just dropped down and then in a matter of seconds it blew up again, and it was an even bigger blast. And all the lights went black; I couldn't see anything anymore. All I could see was fire. When I got on the lifeboat deck, normally everybody was supposed to line up and wait to go on the lifeboat. It was just a complete chaos; people running everywhere. You could see everybody's face. Everybody knew they were going to die. And we knew we were running out of time. It was a matter of time for the derrick to fall and nobody knew where it was going to fall. You could just hear that sound. It was like a jet engine. You can just imagine flame shooting out of a jet engine. Shootin' straight up. This how much pressure the well was blowin'.

Before he died on April 20th, 2010, in the conflagration aboard the Deepwater Horizon, crew member Shane Roshto shared a dark assessment with his wife.

"From Day One he deemed this hole a well from hell," widow Natalie Roshto told an investigating panel Thursday. "He said Mother Nature just doesn't want to be drilled here.[8]"

The remotely operated vehicles determined that the riser pipe disconnected from the rig some time after the explosion. The collapsed riser became a string of tangled spaghetti. This huge, 1 mile long pipe fell down on the seafloor bending severely just above the LMRP, following the seafloor, then rising vertically up 1,500 feet, only to bend sharply downward back to the

---

[8] *Hearings focus on possible human factors in BP oil spill*, by David S. Hilzenrath, Washington Post staff writer, Thursday, July 22, 2010.

seafloor, rupturing on impact, and following the seafloor again. A leak developed at the ruptured connection between the LMRP and the riser. Another major leak developed 600 feet from the BOP. Of course the riser's end and the smaller drill pipe sticking out from it were leaking as well.

The Aftermath in the Gulf

The following is a timeline of the key events that eventually led to plugging the Macondo well and stopping the spill.

- April 20, 2010, at 9:50 p.m.: The Deepwater Horizon drilling rig exploded and caught fire. Eleven crewmen were killed and 115 people were rescued, 17 with serious burns and other injuries.
- April 22, at 5 p.m.: The burned-out rig sank.
- May 2: BP started to drill the first of two relief wells that would intercept the Macondo well sometime in August or September, and be used to fill it with drilling mud and cement.
- May 5: BP inserted a 4-inch pipe into the 6-5/8 inch drill pipe and managed to siphon off some of the leaking oil.
- May 6: A barge arrived at the scene of the Gulf of Mexico oil spill carrying a five-story, 100-ton containment dome ("cofferdam") that BP hoped would contain the leak. This dome was poorly designed, filled with methane ice crystals and floated up one mile to the water surface, nearly colliding with a ship.[9]
- May 9: BP announced that it might be able to plug the well by injecting heavy drilling mud (the "top kill") or shredded tires and golf balls (the "junk shot") below the blowout preventer. Many engineers, including Patzek, publicly discounted the probability of success of the top kill procedure. Unfortunately, they were right.
- May 12: BP lowered a smaller dome to contain one of the leaks from the broken riser pipe.

---

[9] "I said: 'What the hell do you mean you've lost the cofferdam? How did you lose it? Don't give me that!'" Mr. Lynch, a BP vice president and a leader of the effort to kill the well, recalled. "This thing has taken off like a damn balloon. "The last thing you'd want is this thing filled with ice crushing the bottom of the vessel." Clifford Krauss, Henry Fountain and John Broder, "Behind Scenes of Oil Spill Acrimony and Stress," *The New York Times*, 8/27/2010.

- May 14: BP announced that it would insert a 4-inch tube into the leaking 19-1/2 inch riser pipe on the ocean floor. The tube would siphon off some of the leaking oil. By May 17, this pipe was able to carry 1,000 barrels of oil per day to a ship above. By May 20, the rate of oil flow through the pipe was said to have increased to 5,000 barrels per day, but dropped to only 2,000 barrels by the 24th.
- May 17: A ship belonging to the National Institute for Undersea Science and Technology discovered underwater plumes (clouds of tiny dispersed oil droplets) up to 20 miles away from the Macondo wellhead. The largest plume found to date was 90 m thick, 3 miles wide, and 10 miles long.
- May 26: BP attempted the top kill using two 3-inch choke and kill lines connected to the wellhead beneath the blowout preventer. BP tried to fight the equivalent of 10–20 3-inch firehoses blasting oil and gas upwards with two firehoses blasting dense drilling mud downwards. As would be expected, BP failed.[10]
- May 29: BP admitted failure of its top kill and junk shot attempts. The only result was probably increased erosion of the wellbore, and an increased flow rate of oil and gas. However, the two 3-inch lines that were used for the top kill would later be used to flow oil up to the surface.
- June 4: BP's first success was in attaching a "top hat" or "containment cap" to its own riser pipe, which was fastened to the stub of the old riser sawed off just above the blowout preventer (the lower marine riser package, or LMRP, to be exact).
- June 6: The top hat began to siphon off 10,000 barrels of oil per day and millions of cubic feet of gas. With time, BP was able to siphon off up to 25,000 barrels of oil and 50 million cubic feet of gas per day using the top hat and the two 3-inch hoses connected to the choke and kill lines below the blowout preventer. Some oil and plenty of gas were still escaping from the leaky top hat.
- July 7: The installation of a new, larger and tighter fitting containment cap began. The installation involved putting a tight flange on the outlet of the Macondo well's LMRP. A 12 feet pipe ("transition spool") with two flanges was then fastened to the LMRP flange. Finally, a three-ram cap was

---

[10] At one point, technicians said in interviews, a plumbing problem on one of the pump ships forced a delay in the operation. Then a screaming match over the radio between two senior engineers ended in one of them threatening to come over and throw the other overboard. *Ibid.*, see Footnote 5, this chapter.

screwed to the top of the transition spool. The new containment system was 30 feet high and weighed 30 tons. Pressure testing started and confirmed that the cap was working properly.

- July 15: The new cap was shut in and the flow of oil and gas from the Macondo well ceased, 87 days after the blowout.[11]
- August 3: Based on the results of the injectivity test, BP started pumping drilling mud as part of the static kill operations. The purpose was to kill and isolate the well, and to complement the upcoming relief well operations.
- August 9: BP announced that the static kill of the Macondo well was complete and the well was cemented from the top.
- September 3: The original damaged blowout preventer was detached from the wellhead to be replaced with a new, higher-rated blowout preventer in preparation for the final kill of the well.
- September 8: The *Deepwater Horizon Accident Investigation Report* was issued by BP.
- September 15: The relief well drilled by the Development Driller (DD) III rig intercepted the annulus of the Macondo well, and cement was pumped into the annulus on September 17.
- September 20: BP, the federal government scientific team, and the National Incident Commander, concluded that the kill operations successfully sealed the annulus of the Macondo well. The kill was official.
- November 06: The detached old blowout preventer was successfully lifted to the surface.
- January 11, 2011: The National Commission on the BP Deepwater Horizon Oil Spill and Offshore Drilling issues its report to the President: *Deep Water – The Gulf Oil Disaster and the Future of Offshore Drilling.*

Such were the events that emanated from a long string of errors of professional judgment, communication, ethics, and common sense, which played out in an environment too unforgiving to tolerate them. The largest oil and gas spill in U.S. history ended, but not the consequences for livelihoods dependent on the Gulf ecosystem.

Between April 20 and July 15, 2010, the oil gushing from the Macondo well spread over several tens of thousands of square miles of Gulf water, reaching the coasts of Louisiana, Mississippi, Alabama, and Florida. An entire region of

---

[11] Government officials insisted on reopening the well, but luckily the BP engineers prevailed.

Fig. 7.8 Tar and feathers do not mix, especially when it comes to birds. (Image source: conservationreport.files.wordpress.com/2008/05/oil-spill-birds1.jpg)

the United States was under environmental siege. Thousands of birds, turtles, dolphins, and an unknown number of fish and shrimp died. Figure 7.8 is just one tragic and gruesome example. Commercial fishing and shrimping were shut down, and tourists stopped renting hotel rooms along the Gulf coast. Tens of thousands of people lost their livelihoods and incomes, and a whole way of life was demolished. By the time the well was sealed on July 15, 2010, after many trials and errors, at least 1.6 million barrels, or an astonishing 66 million U.S. gallons, of oil gushed into the Gulf waters, and some 800,000 barrels were captured and skimmed.

An Estimated Rate of Oil Production from the BP Well

Because the Macondo well was capped by BP on July 15, 2010, and all flow stopped, we may never know for sure how much oil and gas it produced. The flow rate estimates ranged from 1,000 barrels of oil per day, all the way up to 60,000–150,000 barrels per day, or more. Almost no one paid attention to the natural gas liberated from the oil that contributed between 1/2 and 3/4 of the total flow seen in the video feeds from the remotely operated vehicles (ROVs) monitoring the well. In June 2010, before capping the well, BP was capturing about 15,000–26,000 barrels of oil per day, and some oil (and a lot of gas) was escaping the leaky "top-hat" attached to the blowout preventer.

In mid-June 2010, the video feeds continued to show oil and gas escaping from the containment hat attached to the failed blowout preventer on top of the well. The brown part of the plume consisted of oil droplets, and the white bubbles were gas encapsulated in hydrate ice skins. These ice-gas bubbles eventually dissolved in seawater and never reached the ocean surface. Thus, it is quite important to distinguish between oil and gas in the mixture escaping the well.

The question of the average oil flow rate is not merely academic. Answering it plausibly could be worth a good fraction of a few to a dozen billion dollars that BP may or may not pay in federal penalties for spilling oil into the Gulf of Mexico.

One can perform a sophisticated and tedious analysis of the oil and gas flow rates in the failed BP well, but approximate results can be obtained using much simpler reasoning. Let's start from the facts we know. Early on in the spill, the measured oil pressure in the reservoir was about 816 atmospheres or 12,000 psi. Also early on, the measured pressures at the wellhead and above the blowout preventer were respectively 238 atmospheres/3,500 psi and 180 atmospheres/2,650 psi. The flow length from the reservoir depth to the wellhead depth was about 4,000 m/13,038 feet. The ambient pressure at the ocean floor remained constant, but the wellhead pressure did not, because the flow channels in the BOP would widen with time through erosion, and the reservoir pressure would likely fall due to depletion. The flow length through the blowout preventer and its attachments was roughly 16.5 m/54 feet.

Flow Through the Annulus

Before September 2010, many engineers suspected the annular space of the Macondo well to be the main flow channel. Their views were changed later by the visual evidence from the lifted BOP and casing hanger, presented in Figs. 7.5 and 7.6. However, engineers need to consider all possibilities of a well failure and then rank their probabilities based on other evidence. Because a breach of the annular space definitely is one of the candidate failure modes of the Macondo well, the potential fluid flow through this space must be calculated and compared with all other competing explanations.

Therefore, let's first assume that the oil and gas flowed through the annular space of the wellbore, equivalent to a pipe with the inner diameter of

Table 7.1 The lengths and diameters of parts of the annular space

Item	OD inch	ID inch	Annulus w/7," feet	Annulus w/9-7/8," feet	Annulus width[a], D_H"
9-7/8" Liner	9.875	8.625	2.409	0	1.625
11-7/8" Liner	11.875	10.711	1.956	0	3.711
13-5/8" Liner	13.625	12.375	316	0	5.375
13-5/8" Liner	13.625	12.375	0	1.334	2.500
16" Intermediate casing	16.000	14.850	0	5.926	4.970
22" Wellhead string, upper	22.000	19.500	0	160	9.625

[a] D_H is the hydraulic diameter of the annulus, equal to its width

2.8 inches.[12] To calculate the flow rate of oil we use the classical Weisbach–Darcy equation of pipe flow, in which the pressure drop along a pipe divided by the pipe length is proportional to the square of the total flow rate divided by the pipe diameter to the power 5. The proportionality constant is the friction factor that depends on the roughness of pipe wall, and the fluid velocity, density, and viscosity. One can reasonably assume that the pipe roughness was equivalent to that of commercial steel, and the overall density of the oil and evolved gas mixture was 0.3 of the density of water. With these assumptions, anyone can calculate the oil[13] flow rate at the Macondo well's BOP to be about 18,500 barrels of oil per day,[14] or about five 3-inch firehoses blasting oil.[15] Although the annular space would have been only one of several flow channels,

---

[12] Because flow resistances are in series, the appropriate average is the length-weighted harmonic average of the hydraulic diameters of the annular space segments listed in the last column of Table 7.1. An argument advanced against this averaging relies on the choking effect of pipe collars and the narrowness of parts of the annular space. The pipe collars are very short and their choking effect would be small. They also would be eroded away by the reservoir fluid and sand. Thus, with time, the flow rate through the annular space would tend to increase.

[13] Initially a single-phase supercritical fluid that was neither oil nor gas flowed in the well. The narrowest annular space at the well bottom was exposed to this low-viscosity supercritical fluid that could easily flow through it.

[14] One-half of the total calculated flow rate of 37,000 barrels per day of oil and some gas liberated from the oil, gushing together from the blowout preventer. This calculation is tricky and highly uncertain, because if gas is given enough time to evolve into bubbles and then free gas, there will be not one, but 2.5–3 barrels of gas flowing out of the well for each barrel of oil.

[15] And 5–15 firehoses blowing gas, depending on how fast the gas was liberated.

this discharge rate alone gave BP no chance of stopping the flow with "top-kill" and "junk-shot" procedures that used the equivalent of two firehoses to blast heavy mud and shredded tennis balls into the incoming oil and gas.

Roughly the same volume of oil and gas flowed through the wellbore and the blowout preventer. We use the same Weisbach–Darcy equation to calculate the equivalent pipe diameter of the flow inside the blowout preventer. This calculated diameter was less than 1.5 inch, the blowout preventer was substantially closed or blocked, and it acted as a 1–1.5-inch choke of the oil and gas flow. Again, this is important because all wells in the undepleted oil and gas fields in the Gulf of Mexico flow through chokes.

Using these simple assumptions, we calculate that some 1.6 million barrels, or 66 million gallons, of oil would flow out of the Macondo wellhead if the annular space were breached. Of this total, some 800,000 barrels were siphoned off and skimmed from the water surface as of July 15, when the well was sealed. Under the Clean Water Act, the penalty for spilled oil is either $1,100 per barrel for the plain old sloppy spills, or $4,300 per barrel for the grossly negligent ones.

Flow Through the Production Casing

Another mode of failure of the Macondo well is flow through the breached production casing. Once the blowout preventer was lifted from the seafloor and visually inspected, and an independent analysis of cement strength was performed, this mode of failure became the top candidate to explain the blowout. It also turned out that the partially closed shear ram in the blowout preventer acted as a choke limiting the flow rate a little if the flow were to occur through the annular space and a lot if it occurred through the production casing. This is a classic example of a confusing outcome of engineering analysis, when yet another external factor controls the outcome and confounds conclusions. Life is never simple and dull for engineers.

Figure 7.5 demonstrates that likely most, if not all, of the hydrocarbon flow occurred through the production casing, and little or none through the annular space. Figure 7.6 shows that the blind shear ram was slightly open, with a crimped and partially cut drill pipe stuck in it, as the reservoir oil and gas pushed through the gap. The shear ram was continuously eroded, starting in the early hours of the blowout, and was activated by "hand" with a remotely operated vehicle some 51 hours after the blowout (Fig. 7.9).

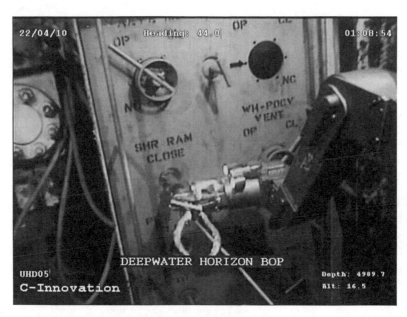

Fig. 7.9 A robotic arm of a Remotely Operated Vehicle (ROV) attempts to activate the Deepwater Horizon Blowout Preventer, on April 22, 2010, 01:08:54 hours, about 2 days after the blowout. When the switch was activated, the BOP stack shook, indicating that the compressed nitrogen cylinders discharged, and the shear ram might have closed. The ram might not have activated automatically upon disconnect from the sinking Deepwater Horizon. For the initial erosion around the ram to occur, the ram must have been partially closed from the beginning

We assume that the bottom 189 feet of the production casing (the so-called "shoe-track") had an effective diameter of 2-3/4 inch, because the unconverted float valves, cement, and debris partially blocked the flow. This diameter had to increase with time to almost 6 inch because of the intensive erosion. With this assumption, the initial equivalent diameter of the casing was 6.9 inches.

To calculate the maximum initial discharge rate under the seafloor conditions, let's also assume that the BOP was wide open. Then the total initial discharge rate would be 253,000 barrels per day, of which 126,500 barrels would be oil. As the cement in the bottom part of the casing would inevitably erode, the total flow rate would increase with time to over 280,000 barrels per day, 140,000 barrels of which would be oil.

This scenario clearly did not happen because we know now that the blowout preventer was significantly closed, as shown in Fig. 7.6. Remember that barring pressure depletion of the reservoir, the Macondo well could deliver over 100,000 barrels of oil per day as its production casing was essentially an

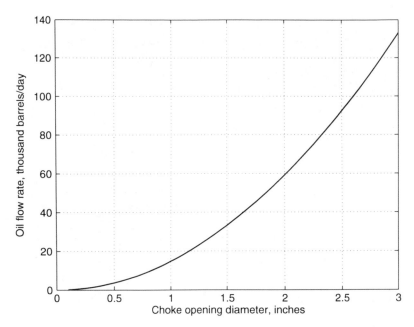

Fig. 7.10 The rate of oil discharge from the Macondo well in thousands of barrels per day, as a function of equivalent diameter of the choke created around the closed shear ram. The oil barrels are referenced to "stock tank conditions" of 14.7 psi and 60 ° F, at which the oil is gas-free. The overall gas–oil ratio of the Macondo crude at seafloor conditions was estimated to be about 3,000 standard cubic feet of gas evolved per 1 stock tank barrel of oil. The flow of oil and gas around the ram is critical, and the 1960 correlation by N. C. J. Ros of Shell was used to calculate the rate. Note that with a choke diameter of 1.25 inch, the oil flow rate is 23,000 barrels per day, similar to the rate calculated above for annular flow. Remember that without a physical examination of the BOP, our calculations are merely informed speculation

unobstructed large pipe straight from the reservoir to the ocean floor. The only major flow restriction was the blowout preventer, but that restriction weakened with time as the wall around the closed blind shear continued to erode.

Assuming that flow channels around the blind shear ram were narrow and small, oil and gas would flow through them at the speed of sound, developing so-called critical flow, whose rate does not depend on the temperature and pressure on the downstream side of the ram, which is open to the ocean. Also assuming that the wellhead pressure remained equal to the 3,500 psi measured by BP early on, one can calculate the total flow rate through the BOP choke as a function of the equivalent diameter of the opening.

The results are shown in Fig. 7.10. If the equivalent diameter of the opening was 1.65 inch, then the flow rate of oil around the shear ram would be 40,000 barrels per day at ocean surface conditions. A 2-inch diameter would result in 60,000 barrels of oil per day, and so on. It is therefore entirely possible that the oil flow rate was initially quite low, perhaps 5,000 barrels per day, and it would gradually increase with time to perhaps as much as 60,000 barrels per day. If the oil rate increased linearly with time, the average rate would be 32,500 barrels of oil per day, and the total amount of oil spilled from the Macondo well would be 2.8 million barrels.

Please remember that after leaving the BOP, each barrel of oil discharging at the seafloor would be accompanied by up to 2.6 barrels of gas, depending on how completely gas could separate. Thus, the video images of the plume taken at the seafloor were dominated by gas bubbles, giving an impression of a much higher flow rate of oil than actually occurred.

The range of oil flow rates obtained here from a simple model of critical flow through a choke agrees well with the estimates provided in the Report by the Deepwater Horizon Spill Team, commissioned by Dr. Marcia McNutt, USGS Director and Science Advisor to the Secretary of the Interior, and Lead to the National Incident Command Flow Rate Technical Group. They mostly evaluated the time-lapse video images[16] of the oil/gas plume emanating from the well, and estimated the discharge rate to be 35,000–45,000 barrels of oil per day, perhaps as high as 60,000 barrels per day.

Inasmuch as we have no access to the BOP, a better estimate will have to wait.

Comparison With Other Wells

The safety track record of the U.S. oil and gas industry has been remarkable: only one major oil spill for all of the 60,000 GOM wells. Compare this with NASA's track record of two lost space shuttles for 132 flights.

Now let's compare the Macondo gusher with other, better-behaved wells in the deep Gulf of Mexico. The BP Thunder Horse platform is the world's largest semisubmersible facility. Prior to the disastrous spill it was also the

---

[16]Their approach is called "Particle Image Velocimetry" or PIV. One of the coauthors, Dr. Paul Bommer of UT Austin, used a classical petroleum engineering approach to calculate the wellbore flow, and the Ros correlation to calculate the choking effect of the BOP. Significantly, for the unobstructed flow through the production casing, Dr. Bommer calculated the oil discharge to be 90,000–105,000 barrels per day.

most productive platform in the Gulf of Mexico, located in water that's about 6,050 feet deep. As of March 20, 2009, daily production at this platform was approximately 260,000 barrels of oil and 210.5 million standard cubic feet of natural gas a day from seven wells. In early 2009, Thunder Horse produced 37,000 barrels of oil and 30 million standard cubic feet of gas per day per average well. The Minerals Management Service reports that the majority of ultra-deepwater wells in the Gulf of Mexico produce around 20,000 barrels of oil a day, with the best well in the entire region producing 41,000 barrels a day from Shell's Mars Yellow Sand. All of these wells are or were constrained by chokes to limit their production rates.[17] The annular space and blowout preventer acted as a choke for the BP Macondo well, whose estimated oil flow rate puts it in the upper echelon of the most productive wells in the Gulf of Mexico.

Next, we look at the confluence of factors that conspired to make the Macondo well tragedy happen. Some of these factors are technical, but most have to do with human behavior, lack of training, poor communication, and plain old confusion.

Further Reading

1. Deepwater horizon accident investigation report, BP. www.bp.com/liveassets/bp_internet/globalbp/globalbp_uk_english/incident_response/STAGING/local_assets/down-loads_pdfs/Deepwater_Horizon_Accident_Investigation_Report.pdf. Accessed 8 Sept 2010
2. Deep water — the gulf oil disaster and the future of offshore drilling, report to the President, National Commission on the BP deepwater horizon oil spill and offshore drilling. www.oilspillcommission.gov/final-report. Accessed 11 Jan 2011
3. Feynman, Richard P.: Report of the Presidential Commission on the space shuttle challenger accident, appendix F — personal observations on the reliability of the shuttle. science.ksc.nasa.gov/shuttle/missions/51-l/docs/rogers-commission/Appendix-F.txt (1986). Accessed 15 March 2011
4. Gardner, Craig, National Commission on the BP deepwater horizon oil spill and offshore drilling — cement testing results, chevron energy technology company. motherjones.com/files/chevron_final_report.pdf. Accessed 26 Oct 2010

---

[17]The completely unrestricted initial production rates of these wells can easily exceed 100,000–200,000 barrels of oil per day. This is why one needs to be careful when drilling any well in ultradeep water in the Gulf of Mexico.

5. Patzek, T.W.: Energy and environment subcommittee of the energy and commerce committee, briefing. democrats.energycommerce.house.gov/documents/20100609/Patzek.Statement.06.09.2010.pdf. Accessed 06 Sept 2010

6. Ros, N.C.J.: An analysis of critical simultaneous gas/liquid flow through a restriction and its application to flowmetering. Appl. Sci. Rev. Sec. A. **9**, 374–388 (1960)

7. Aliseda, A., et al.: Deepwater horizon release estimate of rate by PIV. , www.doi.gov/deepwaterhorizon/loader.cfm?csModule=security/getfile&PageID=68011. Accessed 21 July 2010

8. Det Norske Veritas, final report for United States Department of the Interior Bureau of Ocean Energy Management, Regulation, and Enforcement Washington, DC 20240, Forensic Examination of Deepwater Horizon Blowout Preventer Contract Award No. M10PX0033, Volume I, final report, and Volume II, Appendices, Report No. EP030842. Accessed 20 March 2011

Chapter 8

Why the Gulf Disaster Happened

Transocean and Halliburton's crews finished cementing the Macondo well at 12:40 a.m. on April 20, 2010. At 5 p.m. on the same day, or some 16 hours later, the fateful negative pressure test began.

After completing the cementation job with no lost returns (i.e., no lost circulation), BP and Halliburton declared the job a success. Nathaniel Chaisson, one of Halliburton's crew on the rig, sent an email to another Halliburton engineer, Jesse Gagliano, at 5:45 a.m. saying, "We have completed the job and it went well." He attached a detailed report stating that the job had been "pumped as planned," and that he had seen full returns throughout the process. Just before leaving the rig, a BP drilling engineer, Morel (who designed the Macondo well), emailed the rest of the BP team to say "the Halliburton cement team . . . did a great job."[1]

The following is our account of some of the steps taken by BP and its contractors that cumulatively lead to the tragic blowout on the late evening of April 20, 2010.

The Small and Large Failures

A deepwater exploratory well, such as the BP Mississippi Canyon Block 252-01 ("Macondo") well in the Gulf of Mexico, is drilled in federal waters, more than three nautical miles from the shore.

---

[1] President's National Commission on the BP Deepwater Horizon Oil Spill and Offshore Drilling, 1/11/2011 Report, page 102.

J.A. Tainter and T.W. Patzek, *Drilling Down: The Gulf Oil Debacle and Our Energy Dilemma*, DOI 10.1007/978-1-4419-7677-2_8, © Springer Science+Business Media, LLC 2012

Exploratory wells are drilled to find new oil and gas deposits, whereas development wells are drilled to produce the already discovered and tested deposits. Water depths range from 20 to 400 feet for jackup rigs to 10,000 feet for semisubmersibles and drillships. Before drilling an exploratory well, an operator (such as BP) conducts[2] geophysical studies to identify the most probable locations of hydrocarbon deposits far beneath the ocean floor. The operator then hires a drilling contractor, for example, Transocean, to drill an exploratory well. The operator chooses the well location, prepares a well-drilling plan, and supervises all drilling activities, which may take several months of 24/7 work by a large and variable team of people. The daily cost of this work is $500,000 per day to rent the ship and drilling crew, and proportionally for all other activities. The cost of a single exploration well may be in excess of $100–200 million. There may be junior partners (here Anadarko and Mitsui) to spread the drilling costs, but legal responsibility for the drilling plan, drilling operations, and formation evaluation is with the operator.

In March 2008, BP paid a little over $34 million to the Minerals Management Service (MMS) for an exclusive lease to drill in Mississippi Canyon Block 252, a nine-square-mile plot in the Gulf of Mexico. The drilling plan for the BP Macondo well was submitted to MMS on March 10, 2009, and approved on April 6, 2009. BP started drilling the Macondo well on October 7, 2009, using the Marianas semisubmersible rig, operated by Transocean. This rig was damaged in Hurricane Ida on November 9, 2009. As a result, BP and Transocean replaced the Marianas rig with the Deepwater Horizon. Drilling with the Deepwater Horizon started on February 6, 2010. The Deepwater Horizon was to finish drilling the Macondo well by the end of February 2010, and move on to a new location by March 8, 2010. In reality, the Macondo well took considerably longer than planned to complete. By April 20, 2010, the day of the blowout, the rig was 43 days late for its next drilling location, which may have cost BP as much as $21 million in leasing fees alone. Running late and paying more for the Macondo well, BP and its contractors made several fateful decisions in the days and hours before the blowout.

The well architecture proposed by BP, approved by MMS, and shown in its final form in Figs. 7.2 and 7.3, provided a contiguous space between the central production casing and the outer pipes. It also provided a large-diameter production casing, ready to carry flow oil and gas at very high flow rates.

---

[2] Usually purchases the 3D seismic data from geophysical data acquisition companies or from governments.

Some of the fateful decisions made by BP were described in detail in the letter from Congressmen Henry Waxman (Chairman of the Committee on Energy and Commerce) and Bart Stupak (Chairman of the Subcommittee on Oversight and Investigations) to Mr. Tony Hayward (CEO of BP), dated June 14, 2010. It should be stated for the record that many of the findings of the Waxman letter turned out to be completely irrelevant to this accident. Nevertheless, we believe that BP did not follow the best industry practices outlined by the American Petroleum Institute (API) and the National Oceans Industry Association (NOIA), and that these deviations should be highlighted. In September 2010, BP published its own interpretation of events, and in January 2011, the President's Commission on the spill issued its report. There were also several press investigations, other reports, and leaks that surfaced by January 2011; a few are listed at the end of this chapter.

Well Design

On April 19, 1 day before the blowout, BP installed the final section of steel tubing (central production casing in Fig. 7.3) in the well. BP had two main options: it could lower a full string of casing from the top of the wellhead to the bottom of the well, or it could hang a liner (see Fig. 7.2) from the lower end of the casing already in the well at 17,168 feet and install a "tieback" on top of the liner. The liner–tieback option would have taken extra time and was more expensive, but it would have been safer because it provided one more barrier to the flow of gas up the annular space surrounding these steel tubes. A BP plan review prepared in mid-April recommended against the full string of casing because it would create "an open annulus to the wellhead" and make the seal assembly at the wellhead the "only barrier" to gas flow if the cement job failed. Equally important, a tieback–liner solution would result in a smaller-diameter production casing linking the tieback with the wellhead. Replacing the large 9-7/8 inch pipe with something narrower would cause the future oil and gas production from the Macondo well to be choked to a lower rate.

As explained below, abandoning the hanger option had the deleterious effect of increasing the uncontrolled discharge rate from the well, but only when combined with other missteps in the well cementing and testing procedures. We want to stress that as of this writing we do not yet have all the facts at hand, so we cannot do all the necessary calculations. Therefore, our statements represent our best assessment about what we think happened.

Centralizers

When the production casing was installed, one key challenge was making sure the casing ran down the center of the borehole. As the American Petroleum Institute's recommended practices explain, if the casing is not centered, "It is difficult, if not impossible, to displace mud effectively from the narrow side of the annulus," resulting in flow channeling and a failed cement job.

Halliburton, the contractor hired by BP to cement the well, warned BP that the well could have a "SEVERE[3] gas flow problem," making it essential that the lowest 330 m/1,000 feet or so of production casing are perfectly centered in the borehole, leaving a uniform gap around the pipe. To achieve this complete pipe centering, Halliburton proposed to use 21 centralizers on the final string of casing, instead of only 6 centralizers as calculated by BP using their company software. BP rejected Halliburton's advice to use additional centralizers. In an email on April 16, a BP official involved in the decision explained: "It will take 10 hours to install them. I do not like this." Later that day, another official recognized the risks of proceeding with insufficient centralizers but commented: "Who cares, it's done, end of story, will probably be fine."

Multiple centralizers can be put into a borehole only once without major problems. They inevitably get a liner or casing tube stuck if it must be pulled out of the hole for whatever reason. Drillers do not like to have too many centralizers, and at times they throw the "excess" centralizers overboard, instead of putting them into the hole. It is by no means certain that the six centralizers installed with the bottom section of production casing were insufficient to keep the casing perfectly centered. We will never know. What we know now, however, is that the "missing" (or not) centralizer subs were an example of the broken communication channels within BP and with the suppliers. In any case, we believe that the missing centralizers were not a primary cause of the well failure.

Cement Bond Log

The cement bond log measures the loss of acoustic energy (from tapping a casing wall) as it propagates through the casing; this loss can be theoretically related to the fraction of the casing perimeter covered with cement. The casing

---

[3] The capitalized letters were used by Halliburton.

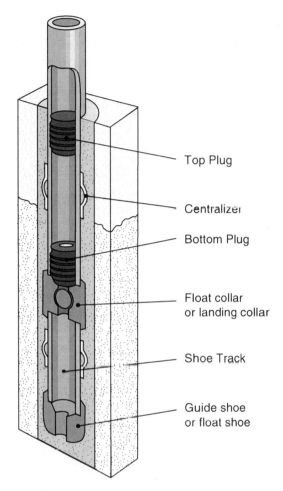

Top Plug

Centralizer

Bottom Plug

Float collar
or landing collar

Shoe Track

Guide shoe
or float shoe

Fig. 8.1 Cementing of the deepest part of production casing in the failed Macondo well. The cement is separated from the drilling mud and a spacer slug by a lower hollow rubber plug that is used to push the mud out from the casing, and avoid cement contamination with the mud that inevitably weakens the cement as shown in Fig. 8.3. A solid upper wiper plug separates the cement slug from the spacer slug and drilling mud above it. The full sequence of operations is described in Table 8.1. The cement flows through a one-way valve out of the casing and into the borehole above. It takes 12–18 hours for the cement to set if there is no contamination. After that period of time a cement bond log should be run to check the quality of cement bond to the outside of the casing pipe and the borehole wall

can be compared with a bell that rings loudly in empty space, but coat it with a layer of cement and you get a dull thud (a nicely cemented pipe rings at less than 5% of the amplitude of a free pipe; see Fig. 8.1).

BP's mid-April plan review predicted cement failure, stating "Cement simulations indicate it is unlikely to be a successful cement job due to formation breakdown (fracturing)." Despite this warning and Halliburton's prediction of severe gas flow problems, BP did not run a 9–12 hours procedure called a cement bond log to assess the integrity of the cement seal. BP had a crew from Schlumberger on the rig on the morning of April 20 for the purpose of running a cement bond log, but they departed at 7:30 a.m., after BP told them their services were not needed.

The problem with the cement bond log (CBL) and foamed cement is that the log might have given erroneous readings, indicating that the well was badly cemented, whereas the cement was fine. In this particular instance, however, the cement fill was probably crushed during the negative pressure test across the bottom 189 feet of the casing, which is inaccessible to the logging tool. Therefore, it is possible that the CBL debate is another moot point, and a diversion of public attention from other, more important failures. On the other hand, performing the CBL would have taken another 12 hours, allowing more time for the cement in the casing to set. It could also tell BP about problems with mixing of cement slugs and instability of foam cement. Perhaps, then, running this potentially useless log would have saved the well. Or, perhaps, it would not.

Mud Circulation and Cement Job

Other missteps were undoubtedly more consequential. In exploratory operations like the Macondo well, wells are generally completely filled with dense mud during the drilling process. The American Petroleum Institute recommends that oil companies fully circulate the drilling mud in the well from the bottom to the top before commencing the cementing process. Mud recirculation sweeps away debris, oil, and gas from the casing interior and the borehole outside the liners. You can think of it as a major cleanup and conditioning of the wellbore that makes it easier for the cement depicted in Fig. 8.1 to bond. Circulating the mud in the Macondo well could have taken as long as 12 hours. BP decided to forgo this safety step and conduct only a partial circulation of the drilling mud before the cement job. Again, completing the circulation would have been nice, but did not cause any disaster. Something altogether different did.

During cementing of the 7-inch casing, a few minor problems were encountered with the "float valves." The bottom 189 feet of the 7-inch casing are called the "shoe track," with a volume of 6.8 barrels (Fig. 8.2). At the top of the

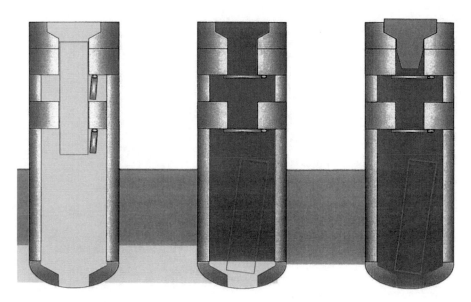

Fig. 8.2 The last segment of the 7-inch production casing is called the "shoe track." In the Macondo well, its length was 189 feet, and volume 6.8 barrels. When the casing is lowered into the hole, the flapper valves at the top of the shoe track are held open by a piece of pipe. The reason is that one wants the drilling mud in the borehole to enter the casing when it is lowered, so that the casing does not get stuck. Prior to the cement job, a ball is dropped onto the pipe to seal it. The pressure inside the casing is then increased to pop the pipe and the ball down and shut the flapper valves, which are spring-loaded. This procedure is called "float conversion." When drilling mud and cement are subsequently pumped into the casing, the flapper valves open under the high pressure. Presumably, 189 feet of neat strong cement were pumped into the shoe track of the Macondo well, the valves should have closed when pressure was lowered and flow stopped, and the casing should have been permanently sealed from the bottom. It would be next to impossible to flow any fluid through the cemented shoe track, if the cement were not contaminated and set properly. But did it? The light color is drilling mud, and the dark one is cement. The shoe track dimensions are not to scale. (Source: BP's May 24, 2010, presentation)

shoe track there are two flapper valves secured in the open position by a piece of pipe. These two flapper valves or "floats" were thought to be "converted" (see the Figs. 8.2 caption) only after the casing pressure increased by more than 3,100 psi, way above the expected increase of 500–700 psi. The pipe that was securing the floats was stuck and the shoe track was probably full of debris. After applying such a high pressure, the debris probably was jetted out of the casing and flowed into the formation below through the "guide shoe"

Table 8.1 Halliburton's plan to cement the casing

Stage	Description[a]
Spacer[b]	72 bbl at 14.3 lb/gal
Plug	Drop bottom hollow rubber wiper plug
Lead	5.26 bbl of premium (Class H) cement at 16.74 lb/gal
Foamed tail[c]	38.90 bbl of premium cement (47.75 bbl when foamed) at 14.5 lb/gal
Shoe track	6.93 bbl of premium cement at 16.74 lb/gal
Plug	Drop top solid closing plug
Spacer	20 bbl at 14.3 lb/gal
Displacement	133 bbl of synthetic oil mud at 14.0 lb/gal
Displacement	728.5 bbl of synthetic oil mud at 14.0 lb/gal, leaving 189 feet of cement in shoe

[a] bbl = barrel = 42 gallons or 5.615 cubic feet; lb/gal = pounds per gallon, seawater has density of 8.6 lb/gal.
[b] A polymer and other additives dissolved in water, used to separate an oil-base mud from a water-base cement that is prone to contamination.
[c] A class H cement foamed with nitrogen. The volume of nitrogen in the cement was $(47.5 - 38.9)/38.9 = 22\%$. By weight, the nitrogen was $(16.74 - 14.5)/16.74 = 13.4\%$ of the cement.

at the bottom of the shoe track. Or possibly the casing parted (i.e., fractured). Either way, the mud started flowing. In order to avoid fracturing the rock around the well, BP decided to go with a low mud circulation rate of up to four barrels per minute, far less than the six barrels per minute needed to unseat a ball in the float assembly and convert the floats. The cementing crew measured a lower-than-expected pressure drop along the casing[4] while circulating mud through it, because – most likely – the floats were not converted and preserved a potential pathway for the flow up from the guide shoe all the way to the wellhead.

But not to worry, according to Halliburton's cementing plan, summarized in Table 8.1, the bottom 189 feet of the casing, or the entire shoe track, would be filled with strong class H cement that could withstand any possible

---

[4] A drilling-mud subcontractor had predicted that it would take a pressure of 570 psi to circulate mud after converting the float valves. Instead, the rig crew reported that circulation pressure was only 340 psi. A BP representative expressed concern about low circulating pressure. He and the Transocean crew switched circulating pumps to see if that made a difference, and eventually concluded that the pressure gauge they had been relying on was broken. Instead, however, the stuck pipe and open floats or the parted casing provided an easier pathway for the mud flow, thus causing a lesser pressure drop.

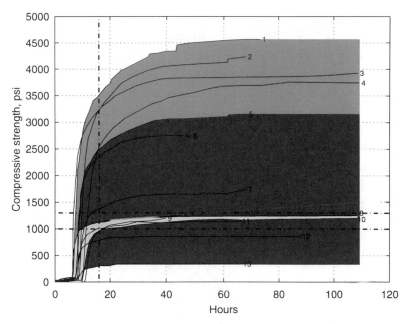

Fig. 8.3 The compressive strength of 1,000 psi means that 1,000 psi of pressure difference across a plug of cement would crush this plug. The three color-filled regions encompass, top-to-bottom, premium H cement with the density of 16.74 pounds per gallon, roughly twice the water density (curves 1–4); premium H cement with different levels of drilling mud contamination (curves 5–7 and 11–13); and premium H cement foamed with nitrogen to the density of 14.25 pounds per gallon (curves 8–10). Drilling mud concentrations of 5% (curve #5), 10% (#6), 15% (#7), 20% (#11), 25% (#13), and 30% (#12) by volume were used to contaminate the neat cement and weaken it very considerably. The foam cement could not be contaminated and tested because addition of the drilling mud would collapse the foam. The vertical dot-dashed line marks time elapsed between pumping cement into the wellbore and the negative pressure test. The two horizontal dot-dash lines denote the increases of compressive stress exerted by the reservoir fluid on the cement, caused by the initial and final stages of the negative pressure test. It is apparent that both the foam cement and the drilling-mud-contaminated neat cement could fail during or after the negative pressure test. Because the flow-blocking float valves apparently did not close, the cement failure would lead directly to a blowout. (Data source: Report, *National Commission on the BP Deepwater Horizon Oil Spill and Offshore Drilling – Cement Testing Results*, Energy Technology Company, Chevron, October 26, 2010)

pressure difference across its column without getting crushed or squeezed out. Or could it? The all too real possibilities of crushing the cement in the shoe track are summarized in Fig. 8.3.

To understand better what Fig. 8.3 means, let us start with the foam cement, a cement slurry puffed up with tiny nitrogen bubbles. If nitrogen is compressed from the surface pressure of 1,000 psi to the downhole pressure of 12,500 psi, and its temperature increases from 60°F to 230°F, the nitrogen's volume decreases about six times. At the downhole conditions, a foam cement is roughly 20% gas by volume, or it has one volume of gas for each five volumes of liquid cement slurry. At the surface mixing conditions, the same slurry will have 1.2 volumes of nitrogen gas for each volume of the liquid. In other words, the foam cement will have more gas than liquid by volume and resemble a low-density froth at the surface mixing conditions. Once this froth is injected into the drill pipe and followed up with a dense neat cement, as in the Halliburton plan in Table 8.1, the dense cement will finger through the froth like a spoonful of raspberry syrup through a foamy beer. If what we are saying is true, there was no class H dense cement filling the shoe track, but only a somewhat altered foam cement. The foam cement has a much lower compressive strength and could be crushed at 1,000–1,200 psi of differential pressure, as depicted in Fig. 8.3.

The foam cement could be crushed if it set at all. But did it? According to the President's Commission report of January 11, 2011, there were several problems with the foam cement (pages 101 and 102): A cement slurry must be tested before it is used in a cement job. Because the pressure and temperature at the bottom of a well can significantly alter the strength and setting rate of a given cement slurry – and because storing cement on a rig can alter its chemical composition over time – companies such as Halliburton normally fly cement samples from the rig back to a laboratory shortly before pumping a job to make sure the cement will work under the conditions in the well. The laboratory conducts a number of tests to evaluate the slurry's viscosity and flow characteristics, the rate at which it will set, and its eventual compressive strength. Chevron's report in the Reading Materials section is a good example of such testing. It appears that the results of Chevron's tests mimic the earlier tests by Halliburton that went unpublished or were improperly transmitted.

When testing a slurry that is foamed with nitrogen, the lab also evaluates the stability of the cement that results. A stable foam slurry retains its bubbles and overall density long enough to allow the cement to set. The result is a hardened cement that has tiny, evenly dispersed, and disconnected nitrogen bubbles throughout. If the foam does not remain stable up until the time the cement sets, the small nitrogen bubbles may coalesce into larger ones, rendering the hardened cement porous and permeable.[5] If the instability is particularly

---

[5] If reservoir fluids start flowing through permeable cement, the cement will be eroded away gradually until it is washed away.

severe, the nitrogen may separate from the cement altogether, forming big bubbles with unpredictable consequences.

On February 10, soon after the Deepwater Horizon began work on the well, BP asked Halliburton to test the cement blend stored on the Deepwater Horizon that Halliburton planned to use at Macondo. They tested the slurry and reported the results to BP. The reported data sent to BP included the results of a single foam stability test that showed that the February foam slurry design was unstable. Documents identified after the blowout revealed that Halliburton personnel had also conducted another foam stability test earlier in February. The earlier test had been conducted under slightly differ-ent conditions than the later one and had failed even more severely. It appears that Halliburton never reported the results of the earlier February test to BP.

Halliburton conducted another round of tests in mid-April, just before pumping the final cement job. By then, the BP team had given Halliburton more accurate information about the temperatures and pressures at the bot-tom of the Macondo well, and Halliburton had progressed further with its cementing plan. Using this information, the laboratory personnel conducted several tests, including a foam stability test, starting on approximately April 13. The first test Halliburton conducted showed once again that the cement slurry would be unstable. The President's Commission did not believe that Halliburton ever reported this information to BP. Instead, it appears that Halliburton personnel subsequently ran a second foam stability test, this time doubling the pretest cement conditioning time to 3 hours. The evidence suggests that Halliburton began the second test at approximately 2:00 a.m. on April 18. That test would normally take 48 hours. Halliburton finished pumping the cement job just before 48 hours would have elapsed, and the apparently positive results of that test had no impact on the Macondo well (the cement bond log [q.v.] might have been useful, but was not performed).

In summary, it is likely that the 7-barrel slug of dense neat cement, meant to fill the casing shoe track, got mixed with the 40-barrel slug of less-dense foamed cement that in turn was likely unstable and might not set properly. The floats did not convert, the possibly improper cement was weak, and all that stood between a safe completion of the well and a disaster was a suffi-cient back-pressure exerted from above by the heavy drilling mud that filled the casing and riser, and the blowout preventer ready to close at a moment's notice. The mud would be removed soon during the butchered negative pres-sure test, and the blowout preventer would not be as full of vigor as most people had hoped.

Lockdown Sleeve

BP did not deploy a casing hanger lockdown sleeve that would prevent the casing from lifting up under pressure and the casing seal from being shredded. BP's final plan was to use the weight of the drill pipe to aid in setting the lockdown sleeve. Setting the lockdown sleeve would require 100,000 pounds of force. The BP Macondo team sought to generate that force by hanging over 3,000 feet of drill pipe below the sleeve. This is the reason for conducting the negative pressure test with over 3,000 feet of drill pipe in the casing, instead of just 1,000 feet or less. Displacing heavy drilling mud with seawater from these 3,000 feet of drill pipe and from the riser, *before* setting a cement plug in the casing, caused the cement in the open shoe track to break and the Macondo well to blow out. BP could have used a shorter length of heavier drill pipe to achieve the same weight on the sleeve, but compromising the well's integrity less.

Thus, it appears that the last-moment ad hoc decisions by BP's management that increased the length of the drill pipe during the negative pressure test, and delayed setting the cement plug until after replacing the drilling mud with seawater, contributed to the blowout.

Negative Pressure Test

The misinterpreted negative pressure test was the last, and perhaps the most important, step that sealed the fate of the well and the rig. Because that test was – and still is – mortally confusing, it needs a more thorough discussion, based on our own calculations.

The negative pressure test relies on decreasing the pressure inside the wellbore by filling all fluid flow pipes above the end of the drill pipe with seawater and waiting for some time. As illustrated in Fig. 8.4, if no extra pressure is measured at the end of each pipe on the rig's floor, and there is no significant water flow out of any of the pipes, the well is deemed sealed. If one or more pipes shows a positive pressure reading, there is a leak in the production casing (central tube) and/or in the wellhead seal separating the annular space from the casing. The central drill pipe showed a very high pressure reading (1,400 psi), but the interpretation of the test was confused by different volumes of much denser muds in different pipes. In the end, BP incorrectly assumed that the test results affirmed that there was no leak in the well. A few hours later, Mother Nature reminded everyone that what you

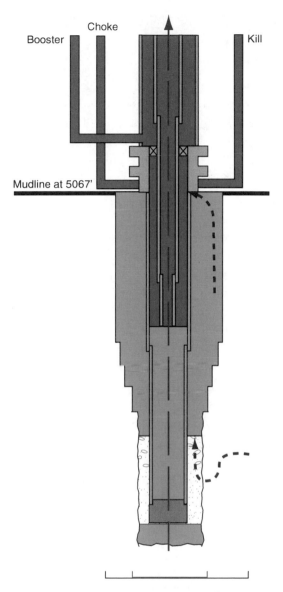

Booster   Choke   Kill

Mudline at 5067'

Fig. 8.4 To test the seal integrity, all pipes above the end of the drill pipe should be filled with seawater, and all the surface pressure readings of the pipe pressure should be zero. They were not; the pressure in the central drill pipe was measured at 1,400 psi, indicating a leak, but the test interpretation was obfuscated by the presence of large slugs of the dense drilling mud, and even denser lost circulation mud "spacer" in the different pipes. At one point of time or another, these mud slugs were present in the kill line, the riser, the drill pipe, and the uppermost 3,000 feet of the production casing. Having three different fluids in four pipes completely confused results of the test. The most likely flow path of reservoir fluid was through the production casing. (Data source: BP, Washington Briefing, Deepwater Horizon Interim Accident Investigation, May 24, 2010, Patzek's calculations)

see on the computer screen, and how you think about it, may disagree with reality. The rig exploded.

So what really happened beginning at noon and ending at 9:52 p.m. on April 20, 2010?

1. At 1:28 p.m., BP started transferring mud from the well to a boat moored adjacent to the rig. For the next 4 hours, until 5:30 p.m., the mud transfer procedure impaired the monitoring of mud appearance and level in the mud pit on the drilling floor. Such monitoring would have told BP if the well started to flow and/or if there were chunks of cement coming out from the wellbore.

2. At 3:03–3:54 p.m., the mud booster, choke, and kill lines[6] were filled with seawater to displace the drilling mud, and the valves that cut off these three lines from discharging into the well were closed. Somehow, a high pressure from the wellbore was transmitted[7] to the kill line, and 1,200 psi was read at its shut-in outlet on the rig floor. The riser, drill pipe, and production casing were still filled with the drilling mud at 14 pounds per gallon, or 1.67 kg/L, about 63% denser than seawater.

3. Now comes the fatal misstep. From 3:56 to 4:28 p.m., 454 barrels of lost circulation mud (LCM) were injected through the drill pipe into the well. LCM is a viscous mud, 100–200 times more viscous than water. The LCM density is 16 pounds per gallon, 1.87 times denser than seawater. It is used in case of emergency, to plug-in the weak rock fractured during drilling. The reason for the injection of these 454 barrels of LCM was as surreal as it was explainable. According to the EPA regulations, LCM stored on the rig is a toxic substance that needs to be disposed of accordingly. But if the same mud is circulated through the wellbore, it can be dumped directly into the sea. The drilling crew did not want to dispose

---

[6] Choke and kill lines are the high-pressure tubes for circulating drilling mud to the subsea BOP. The mud booster line provides additional mud to the base of the riser to maintain return fluid velocity.

[7] The riser was filled with drilling mud, and at the wellhead, it exerted the hydrostatic pressure of 3,688 psi. The kill line was now filled with seawater, and it exerted the pressure of 2,258 psi at the wellhead. The pressure difference, 1,431 psi, had to be applied at the rig, to push the seawater into the kill line. Somehow, most of this extra pressure was not bled off from the kill line when its bottom valve was closed. Why this was not done remains a mystery; maybe the bleed-off valve was stuck, and the people in charge were rushing.

of the toxic chemicals, so they injected the fatal 454 barrels of LCM into the well to circulate them back through the riser and pump overboard once it passed a cursory "sheen test" for the presence of crude oil. This volume of LCM displaced all drilling mud from the drill pipe and from the production casing down to 8,367 feet. LCM then spilled over into the annular space of the BlowOut pressure Preventer (BOP), filling 54 feet of the BOP and 216 feet of the riser pipe above the BOP, to the depth of 4,797 feet. Now we had two different fluids in the wellbore (drilling mud and LCM), and three different fluids above the wellhead: sea water in the choke, kill, and booster lines, LCM in the drill pipe, and LCM in the BOP and riser up to 270 feet above the wellhead, with ordinary drilling mud above it all the way up to the rig. It was very difficult to interpret pressure readings when three different fluids started flowing in four pipes and without exact knowledge of where these fluids went and how they mixed.

4. It seems that the crew was expecting gas in the pressured-up kill line, but none flowed out of the kill line. The crew thought that the bleeder valve was functioning properly, but probably it was not.

5. Between 4:29 and 4:52 p.m., 352 barrels of seawater were injected at the rig into the drill pipe. In a perfect world, this amount of water would have displaced 352 barrels of LCM, and left 21.4 barrels of LCM in the wellbore below the wellhead. The 54 feet of the open BOP and 1,332 feet of riser pipe (up to a depth of 3,681 feet) would have been filled with LCM. In order to keep this height of mud in the riser, the extra pressure applied to the top of the drill pipe would have been 1,767 psi. The measured pressure at the top of the drill pipe was 2,325 psi, 559 psi higher. Something else was going on. The heavy viscous LCM mud was not displaced from the production casing by the lighter thin water. Instead, the injected seawater fingered through the mud and mixed with it. In order to obtain the pressure of 2,325 psi at the top of the drill pipe, one can calculate that only 272 barrels of the original 454 barrels of LCM were displaced and the remainder mixed with water forming 282 barrels of a lighter mud with the density of 13.4 pounds per gallon. The rest of the injected water filled the drill pipe.

6. Between 4:53 and 4:55 p.m., the crew shut in the lower annular, thus, they thought, isolating the wellbore and drill pipe from the contents of the riser.

7. Between 4:55 and 4:57 p.m., the crew bled the drill pipe down to 1,220 psi, or by 1,105 psi.

8. Between 4:57 and 5:05 p.m, the crew opened the kill line valve and the LCM/seawater mixture flowed into it. Pressures in the drill pipe and kill line equalized at 1,220 psi for a short while, then the drill pipe pressure increased to 1,400 psi and the kill line pressure decreased to 645 psi. The pressure in the drill pipe was bled to 273 psi, and the pressure in the kill line went to zero. The drill pipe pressure never went to zero, indicating continuing fluid flow, but was interpreted by the crew as a problem with the pressure gauge. It was observed that the mud level in the riser was dropping. There was flow out of the riser, whereas fluid had to be flowing out of the drill pipe and/or kill line. Obviously, the closed-off annular was leaking. The kill line was now filled with 44 barrels of LCM/water mixture, whose density was about 13 pounds per gallon. This dense mud filling the kill line in place of sea water imposed a 60% higher hydrostatic pressure at the bottom of the kill line. Therefore, the pressure readings at the top of the kill line were 60% lower than expected and made the crew believe that there were no problems with the well. The drillers decided to trust the kill line pressure gauge more than the troublesome high-pressure readings from the drill pipe gauge that was filled with seawater.

9. Between 5:05 and 5:25 p.m., the drill pipe was shut in, allowing pressure in it to build to 1,250 psi in 6 min. The riser was refilled with 50 barrels of mud, and the mud level was steady. The drillers concluded that the annular was not leaking now, whereas in reality the mud weight could have been supported by the increasing wellhead pressure. The decision was made to conduct a negative pressure test, that is, to circulate LCM and drilling mud out of the drill pipe, some of the production casing, and out of the riser. This procedure would lower the wellhead pressure by more than 1,000 psi, and allow the casing shoe cement to break. See the light color band in Fig. 8.3.

   To break the narrative for a moment, the drilling crew was dealing with a hopelessly complex situation, and no one quite understood what was going on. Because they all wanted to leave the site, they decided to proceed with the negative pressure test, not knowing fully what the fluid distribution and the wellhead pressure were.

10. Between 5:27 and 5:52 p.m., the drill pipe was bled from 1,202 to 0 psi. At that point the drill pipe was closed off with an Internal BlowOut Preventer (IBOP). Before closing the IBOP, the drill pipe discharged 15 barrels of seawater, instead of the expected 5 barrels, indicating flow in the well for the first time. The rig still had 4 hours to escape its fate.

11. Between 5:52 and 6:40 p.m., the crew was watching the well. The pressure in the drill pipe quickly rose to 790 psi and then fell. The crew now opened the kill line and bled 3–15 barrels (up to one third of the kill line's volume). A witness reported that the well was flowing and spurted, indicating gas or heavy viscous drilling mud in the line. They shut in the kill line and saw its pressure build up.

12. Between 6:40 and 7:55 p.m., the drill pipe pressure increased back to 1,400 psi. The crew applied vacuum to the kill line to make sure it was full, and bled off 0.2 barrels. They monitored the kill line for 30 minutes and prepared to displace the contents of the drill pipe and riser with seawater. With 1,400 psi on the drill pipe the crew decided that the test was successful.

    Again, we break the narrative. The crew on the rig watched the well for 2 hours, saw it flowing, saw the pressure increasing to 1,400 psi, and decided that everything was fine with the well. To make things worse, an onshore team of specialists from BP went home after 5 p.m. They were supposed to monitor via satellite links what was happening on the rig and evaluate the negative pressure test in real-time. Despite all of the flat screens in the room, full of colorful plots and flashy displays, no one was watching or understood the true meaning of the late evening news from the rig flashing on the lonely unattended monitors.

    So now, as in a Greek tragedy, the audience already knows the outcome, but the actors on the scene are oblivious to the events that are unfolding in front of their wide shut eyes.

13. Here comes the coup de grâce. Between 7:55 and 9:14 p.m., the crew pumped 1,304 barrels of water through the drill pipe into the wellbore, thus displacing most of the 454 barrels of LCM from the riser. Remember, the volume of annular space in the riser was 1,650 barrels (Table 8.2), and the crew injected $1,304 + 352 = 1,656$ barrels. They shut down the pump to look at the returned LCM in a tank, and check that it was not contaminated with oil. They were satisfied that LCM passed the sheen test and was ready to be pumped overboard. When the LCM slug was circulated out, the pressure reading on the drill pipe was 1,017 psi, but it should have been much closer to zero. The well was clearly leaking. The flow out of the well was larger than the flow in at 8:58 p.m. At that time the pumps were slowed down and the drill pipe pressure increased from 1,250 to 1,350 psi. The well flowed out a net 57 barrels over 12 minutes. Ten minutes later, at 9:08 p.m, the pump was shut down for 5.5 minutes to observe LCM for the presence of oil, but the well continued to flow, and the drill pipe pressure continued to increase.

Table 8.2 Dimensions and volumes of well tubulars

Item	OD (inch)	ID (inch)	Top (feet)	Bottom (feet)	Volume (barrels)
Kill line	3-7/8	3.000	0	5,067	44.0
Drill pipe	6-5/8	5.761	−30[a]	7,545	244.2
EUE[b] stinger	3-1/2	2.992	7,545	8,367	7.1
Drill pipe	–	–	–	–	251.4
Riser pipe	21-1/2	18.750	0	5,067[c]	1730.5
Casing tube	9-7/8	8.625	5,067	12,487	535.6
Casing tube	7	6.094	12,487	18,304	210
Riser annular	–	–	–	–	1514.5
Casing annular[d]	–	–	–	–	121.8

[a] 30 feet above the Rig's Kelly Bushing (RKB). The kelly bushing connects the kelly to the rotary table. The hexagonal kelly turns the entire drillstring.
[b] EUE stands for the "externally upset end." The pipe has an eight round thread connection (eight threads per inch with a rounded thread profile).
[c] The blowout preventer and riser have the same diameter, so we go down to the ocean floor (mudline).
[d] Production casing to the depth of 8,367 feet.

The well was clearly out of control and flowing. The rig team was tired and confused, and the onshore team went home. Instead of immediately pumping LCM or drilling mud into the drill and/or kill pipes, and/or shutting in the BOP, the crew was looking for oil sheen on the mud, which to begin with they probably should never have injected into the wellbore.

14. Instead of sailing away as fast as they could, the crew proceeded calmly to dump LCM overboard, and pumped another 256 barrels of seawater into the drill pipe. The flow meter was bypassed, so no one watched the flow out. At 9:47 p.m., the drill pipe pressure increased to 5,700 psi.

15. At 9:49 p.m., there was the first explosion, and 9:56 p.m., the rig captain ordered EDS, the emergency disconnect sequence. It was already much too late.

After the explosions, the ship lost power and started drifting. The electronic communication lines connecting the ship and the blowout preventer were most likely severed instantaneously, but the blind shear did not fully deploy, presumably because of the bent, off-center drill pipe in it. The drifting ship probably started dragging the still-attached drill pipe through the blowout preventer, damaging the annular preventers and bending the pipe. The pipe then broke off from the ship and fell down with the riser onto the seafloor, contorting severely. Some 50 hours later the shear ram was activated by a remotely operated vehicle, and almost closed, but the erosional damage of the ram shears had already started, allowing for a relatively small leak that would grow with time.

How Could This Happen?

In summary, the conjunction "and" helps to explain the domino effect of interrelated events.

1. BP wanted to drill the Macondo well to the depth of 20,200 feet, but could only drill to 18,300 feet, because the deepest formations were too fragile and too difficult to drill.
2. And, BP did not know this situation ahead of time, because the Macondo was their first exploratory well in the Mississippi Canyon Block 252.
3. And, because the deep rock was too close to being fractured, BP had to use a low mud circulation rate and a foamed cement.
4. And, because the circulation rate was too low and the floats were rate-sensitive, they were not converted.
5. And, because BP used a likely unstable and weak cement designed by Halliburton, and did not convert the floats, an incipient flow pathway was created to the seafloor.
6. And, because BP did not perform the Cement Bond Log, they accelerated the negative pressure test by at least 12 hours, and did not learn about the likely serious problems with the foam cement.
7. And, because BP wanted to save time and decrease risk when setting the casing lockdown sleeve, they used over 3,000 feet of drill pipe below the seafloor, and by displacing mud from the pipe with seawater caused the well to be more underbalanced than it would have been otherwise.
8. And, because BP wanted to set the lockdown sleeve first and a cement plug later, the incipient flow pathway remained open.
9. And, because BP, Halliburton, and Transocean misinterpreted the negative pressure test, that pathway opened and became all too real.
10. And, the blowout preventer did not fully close.
11. And, last but not least, the Minerals Management Service gave its blessings to it all.

And there was a blowout. And 11 people died. And a whole way of life was ruined for a long time.

What Went Wrong with Management?

In this book we distinguish between local and systemic reasons for the major technology-related failures. First, let us discuss the local causes for the Macondo tragedy.

Initial design of an offshore well undergoes serious scrutiny and review by peers and regulators. Inevitably, this well design is modified during drilling and completing the well. Difficulties and problems encountered while drilling a well are probably the main reason for the modifications. Changes to drilling procedures in the weeks and days before implementation are typically not subject to a comparable peer review. At Macondo, such late decisions appear to have been made by BP's on-site management team in an *ad hoc* fashion with no formal risk analysis or internal expert review. The lax and capricious approach of the local BP management seems to have significantly contributed to the blowout.

The Halliburton and BP managers did not ensure that the all-important nitrified cement, whose failure doomed the ship, was adequately tested. As observed in the Presidential Commission's Report, Halliburton had insufficient controls in place to ensure that laboratory testing was performed in a timely fashion or that test results were vetted rigorously in-house or with BP. It now appears that Halliburton did not complete the essential tests of the Macondo cement slurry stability until after the cement had been pumped. It is difficult to imagine a clearer failure of management and communication.

BP, Transocean, and Halliburton failed to communicate adequately. Information appears to have been excessively compartmentalized at Macondo as a result of poor communication. BP did not share important information with its contractors and sometimes internally with members of its own team. Contractors did not share important information with BP or each other. As a result, individuals often found themselves making critical decisions without a full appreciation for the context in which they were being made. Some of these individuals may have been exhausted, distracted, inattentive, or incompetent (although this is merely to observe that they were a normal cross-section of humanity).

Then, as pointed out by Patzek in his congressional testimony of June 9, 2010, there have been strategic, equally unsettling, trends in the development of offshore technology and related human resources:

1. The federal government has virtually abandoned all offshore technology-related research, and the oil and gas industry has eliminated most of its research capabilities, which three decades ago allowed it to rapidly expand deepwater production.

2. Large service companies have been unable to satisfy the ever-growing research needs of the industry.

3. Academic research has been important but small in scale and permanently starved of funding. Within academia, petroleum engineering departments have generally been viewed as less important and glamorous than, for example, biomedical departments. This attitude has resulted in an almost uniform understaffing of petroleum engineering departments in the United States.

4. The depletion of industry research capabilities, and the starvation of academia that educates the new industry leaders,[8] have resulted in a scarcity of experienced personnel that can grasp the complexity of offshore operations and make quick and correct decisions.

5. The industry has replaced the educated knowledgeable people with software that is usually written by programmers with computer science degrees, but with little knowledge of the domain physics. This increasingly complex software often gives answers that are difficult to interpret or plainly false.

6. To make things worse, a vast majority of the current industry engineers and scientists are above 50 years of age and will retire soon.

7. Oil companies no longer have sufficient in-house manpower that would allow them to be unequivocally in charge of complex offshore operations. Instead, they must rely on multiple contractors, who independently perform the various tasks related to drilling and completing a deepwater well. The individual contractors have different cultures and management structures, leading easily to conflicts of interest, confusion, lack of coordination, and severely slowed decision making. The lack of a clear line of authority is particularly damaging in extreme situations, such as the Deepwater Horizon explosion and sinking.

8. Similar observations apply to the government agencies involved in spill management. They suffer from extreme bureaucracy, incompetence, overlapping authorities, and the absence of clearly delineated chains of command.

NASA and the space program offer us some illuminating comparisons. Most people over the age of 30 will have a vivid recollection of the explosion of the Space Shuttle *Challenger* over Cape Canaveral on January 28, 1986

---

[8] Jon Stewart asked on his show, "Anybody here majored in oil leaks? Anybody? Anybody?"

Fig. 8.5 The Space Shuttle *Challenger* disaster occurred on January 28, 1986, when the shuttle broke apart 73 seconds into its flight, leading to the deaths of the seven crew members. The spacecraft disintegrated over the Atlantic Ocean, off the coast of central Florida, United States. Disintegration of the entire vehicle began after an O-ring seal in its right solid rocket booster failed at liftoff (Image sources: NASA and Wikipedia). Physicist Richard Feynman with an O-ring in a G-clamp is shown in the *upper left*; the *Challenger* explosion is in the *center*; and a fragment of the failed rubber O-ring and booster, both lifted from the seabed, are shown in the *lower right*

(Fig. 8.5). As part of the investigation of the root cause of this disaster, the legendary physicist and Nobel Prize winner, Richard Feynman, insisted on expressing some unpopular views in writing. The Air Force brass and government bureaucrats would have none of it, so Feynman's thoughts were relegated to Appendix F — *Personal Observations on Reliability of Shuttle* of the rather dull *Report of the Presidential Commission on the Space Shuttle Challenger Accident*, issued in June of 1986.

Appendix F of the Report is a priceless record of a way of thinking that (if ever applied in practice) would prevent most future disasters, the Macondo well included. We have bracketed the words "[vehicle]," "[Shuttle]," and the like from the most applicable statements by Feynman, so that you can insert

Table 8.3 The simple failures of complex systems

Complex system	System cost	Failed part cost
Space shuttle	$1.7–6.7 billion	$1,000 [a]?
Thunder horse	$1 (+1) billion	$100 [b]?
Deepwater horizon	$700 [c] million (+30 billion?)	$20 million [d]?

[a] Failed O-ring was a fluoroelastomer specified by Morton-Thiokol.
[b] A 6-inch length pipe (but also bad welds).
[c] $500 million for the rig and $200 million for the well with cost overruns.
[d] Pull BOP and check it, properly test cement, run CBL, put in cement plug before negative pressure test, and avoid 3,300 feet of tubing below wellhead to lock down casing.

your own phrases: "Macondo well," "blowout," "drill," "cement," and so on. Prophetically, Feynman said:

> It appears that there are enormous differences of opinion as to the probability of a failure with loss of [vehicle] and of human life. The estimates range from roughly 1 in 100 to 1 in 100,000. The higher figures come from the working engineers, and the very low figures from management. What are the causes and consequences of this lack of agreement? Since 1 part in 100,000 would imply that one could put a [Shuttle up] each day for 300 years expecting to lose only one, we could properly ask "What is the cause of management's fantastic faith in the machinery?" We have also found that certification criteria used in [Flight Readiness] Reviews often develop a gradually decreasing strictness. The argument that the same risk was [flown] before without failure is often accepted as an argument for the safety of accepting it again. Because of this, obvious weaknesses are accepted again and again, sometimes without a sufficiently serious attempt to remedy them, or to delay a [flight] because of their continued presence.

It is surprising to compare the cost of a failed part or bad procedure with the total loss of life and money in a major disaster. The costs of simplistic thinking about three complex systems are listed in Table 8.3, and the most important conclusion of this chapter is: a complex multibillion dollar system disintegrates because of one or few poorly designed parts, or procedures, that cost almost nothing. Bad management, judgment, and workmanship are always involved, and almost impossible to prevent.

As noted elsewhere, if a problem has more than one superlative, the problem itself becomes completely meaningless and indeterminate. George Kingsley Zipf wrote about it in his masterpiece, *Human Behavior and the Principle of Least Effort — An Introduction to Human Ecology*, 60 years ago. Before him it was pointed out that the frequent statement, "In democracy we believe in the greatest good for the greatest number," contains two superlatives and therefore is meaningless and indeterminate.[9] The singleness of objective follows exactly the same principle. The concept of singleness of superlative or objective applies to both human activities and ecosystems in biology. BP management obviously did not read Zipf, and confused all BP employees and all who worked as BP contractors. Lord John Browne and Tony Hayward, the previous two CEOs of BP, failed to follow this most basic principle of management. The results were tragic for BP and caused a string of accidents, safety violations, and deaths that finally ushered in the Macondo well accident.

Many people claim that real-world decision-making problems are usually too complicated to be solved with use of a single objective of arriving at the "best" solution (criterion of optimality). They also claim that such an unsophisticated approach can lead to unrealistic decisions. However, the practical multiple-criteria decision-making process by BP and its contractors led to a verifiably nonoptimal result: They aimed to save a few tens of millions of dollars, but these potential savings would cost them many tens of thousands of millions of dollars, not to mention the loss of happiness by many millions of Gulf coast inhabitants, including the turtles and pelicans. What is the fair monetary value of a lost way of life and lost happiness?

We have an understandable need to blame someone and assign responsibility for the disaster, and the leadership and public face of BP are our obvious choice. Yet responsibility is different from cause, and there were factors at work that transcend the culture and competence of a single corporation and/or its drilling and cementation subcontractors. Part of the blame must rest with the rubber stamp of the U.S. government regulators for such a technologically complex and risky operation, and then there is ourselves as unwitting participants in an energy–complexity spiral that encourages and rewards high levels of risk taking.

Cheap abundant energy, chiefly from oil, has come to be regarded as an American birthright, and we all expect someone to drill and deliver that oil to support our energy-exuberant lifestyles. The tragedy aboard the Deepwater Horizon may be a rare event, like a Black Swan, but it does force us to consider the potential price for the complex and risky technological solutions that will continue to be required to bring the remaining oil to market.

---

[9] Here is a good example of confusion: "Safety was a bigger component than downtime," Mr. Winslow [a Transocean employee], said. "Safety is one of our core values." Adviser says he raised concerns to BP on well, Robbie Brown, *The New York Times*, 8/25/2010.

Summary and Conclusions

In this chapter we have described what we see as the key reasons for the Macondo well accident. Failures of judgment, communication, and execution were the root causes of this tragedy. These breakdowns of performance are shared by BP, Halliburton, and Transocean, who all made a mutually interlocking series of missteps, in which mistakes by one party would amplify the mistakes made by other parties to the tragedy. The conjunction "and" was used to paint a full picture of the interrelated failures.

In their postaccident presentation, BP had shown eight different failure modes that had to align to cause the Macondo well blowout. These failure modes aligned because of the repeated human errors and inaction, or delayed action. Essentially, on top of the serious flaws of well design, nobody detected the well failure until it was too late. The human errors might be summarized as an overreliance on the tired, confused, and distracted personnel, inadequate and often-changing instructions, inadequate training and procedures to stem emergencies, and inattentive management or rig's crew. A dire lack of communication within and among the different companies also played a role.

We hope that readers have gained an understanding of the complexity and perilousness of offshore well-drilling operations. Many things can go wrong when drilling a well in ultra-deep water, and sometimes they do. First, one needs to have a well design and drilling plan that accounts for unplanned contingencies,[10] and provides for at least two barriers to fluid flow at any stage of well-drilling operations. The blowout preventer should be treated as an additional or backup barrier. Second, the key to avoiding confusion and controlling danger is to have a clear work plan that is guided by a single imperative: safety. Well-drilling operations should be stoppable by anyone on the ship, or onshore, who has a valid reason to do so. This singleness of imperative (or superlative) was recognized by cognitive scientists 70–80 years ago. A job description becomes confusing and meaningless if a company says that safety is important and the lowest costs are important and the fastest turnaround time is important. The reason for the singleness of imperative is physical: no problem in dynamics can be properly formulated in terms of more than one imperative (or superlative), whether this superlative is stated as a maximum (of safety) or a minimum (of cost).

---

[10] We know how well our government handles the unknown unknowns, Mr. Rumsfeld's self-congratulations notwithstanding.

Here is another example. Who was in charge of the rig, the captain or the offshore installation manager? The work crew has given conflicting opinions about which of the two officials was in command of the rig during the final emergency. As Captain Hung Nguyen of the U.S. Coast Guard correctly observed: "Everybody in charge, nobody in charge."

Modern offshore rigs and production platforms are often large complex facilities, operated by 120–160 people at any given time. These people must be trained and managed properly, and they need to communicate and work together. The root cause of the Macondo well blowout was failure of management, training, and communication.

Further Reading

1. Bulgakov, M.: The Master and Margarita. Grove Press, (1988): Probably the best novel ever written, with a sweeping treatment of good, evil, and punishment
2. Deepwater Horizon Accident Investigation Report, BP. http://www.bp.com/liveassets/bp_internet/globalbp/globalbp_uk_english/incident_response/STAGING/local_assets/downloads_pdfs/Deepwater_Horizon_Accident_Investigation_Report.pdf. (2010) Accessed 9 August, 2011
3. Deep Water – The Gulf Oil Disaster and the Future of Offshore Drilling, Report to the President, National Commission on the BP Deepwater Horizon Oil Spill and Offshore Drilling. http://www.oilspillcommission.gov/final-report (2011) Accessed 9 August 2011
4. Gardner, C.: National Commission on the BP Deepwater Horizon Oil Spill and Offshore Drilling – Cement Testing Results, Chevron Energy Technology Company. http://motherjones.com/files/chevron_final_report.pdf. (2010) Accessed 9 August 2011
5. Patzek, T.W.: Self-similar collapse of stationary bulk foams. AIChE J. 39(10), 1697–1707 (1993). gaia.pge.utexas.edu/papers/SelfSimilarCollapseOfFoams.pdf: A lot of physics and math pertinent to foam collapse
6. Perrow, C.: Normal Accidents: Living with High-Risk Technologies. Princeton University Press, Princeton (1999). Accidents are unavoidable, and no safety precautions will reduce their risk to zero. http://www.oilspillcommission.gov/final-report (2011). Accessed 9 August 2011
7. Zipf, G.K.: Human Behavior and the Principle of Least Effort – An Introduction to Human Ecology. Hafner Publishing Company, New York (1949). A classical discourse of how human communication works. http://www.oilspillcommission.gov/final-report (2011). Accessed 9 August 2011

Chapter 9

Our Energy and Complexity Dilemma: Prospects for the Future

If fish were scientists, suggests our colleague T. F. H. Allen, the last thing they would discover would be water. We are not sure where this saying originates. Something like it appeared in *The New York Times* in 1920 in a report on a talk by British scientist Sir Oliver Lodge. "Imagine a deep-sea fish at the bottom of the ocean," lectured Sir Oliver. "It is surrounded by water; it lives in water; it breathes water. Now, what is the last thing that fish would discover? I am inclined to believe that the last thing the fish would be aware of would be water."[1] We like a variant of this conundrum: imagine that you could talk to a fish, and ask the question: Is your nose wet?

The fish-discovering-water dilemma illustrates why the Gulf oil spill came as such a surprise, why it appeared to be a Black Swan. Fish are unlikely to discover water because it is the context in which they are immersed. Evidence of water (due to suspended oil droplets, perhaps) would be the fish's Black Swan, something never before perceived. Of course the water had been there all along. Water for fish is like air for us. There is an old joke in which a resident of Los Angeles asks a rural person, "Doesn't it frighten you, breathing something you can't see?" Air is part of our context, so we notice it only when something is wrong, when it is no longer invisible.

Although any organism's context includes large-scale intangibles such as water or air, humans have a context that is peculiar to us, similarly intangible, and special. That context is our culture. Aside from fanciful stories of children reared by wolves, each of us is raised in a cultural context that shapes

---

[1] "Lodge Pays Tribute to Einstein Theory." *The New York Times*, February 9, 1920.

J.A. Tainter and T.W. Patzek, *Drilling Down: The Gulf Oil Debacle and Our Energy Dilemma*, DOI 10.1007/978-1-4419-7677-2_9, © Springer Science+Business Media, LLC 2012

much of the person we become. This is the process called socialization. Through socialization, a child is molded to become a functioning member of a particular society—American, say, rather than Chinese. Some of this happens through formal training by both parents and educators, but much of it is informal and below the level of conscious recognition. Children observe their parents, other adults, siblings, and peers (and today, social media, television, and even virtual videogame personalities) thinking and behaving in patterned ways, and unconsciously adopt those thoughts and behaviors themselves. Because this happens at a very early age, culturally pre-scribed thoughts and behaviors appear to be natural. A young child ordinarily experiences no alternative patterns of thought and behavior, so the context in which he or she is socialized becomes normal and unexceptional. Moreover, ongoing brain research is showing that early socialization and experience cause neuronal connections in a child's brain to form in certain ways. Thus we are all, in this sense, hard-wired. Most people, later in life, never question the cultural context in which they were raised: it is natural, and even comfort-able. When encountering members of other cultures, most of us will find their thoughts and behaviors odd, bordering on unnatural, and possibly a little unsettling or even threatening. Most of the time culture is our unper-ceived context, like air or water.

Certain kinds of gradients are also part of our cultural context, unper-ceived like air or water. Nature abhors both vacuums and gradients. The essence of a vacuum is that it is a space with no pressure. Another way of putting this is that a vacuum is a space with the greatest difference in pressure between the inside (with theoretically no pressure) and the outside. In this respect, a vacuum is an extreme kind of gradient. Commonly we think of a gradient as a slope or incline. A hill or grade (as in gradient) is a common example (Fig. 9.1), but the term can also refer to the difference in magnitude of wealth between a first-world country and a third-world neighbor. It can refer to the potential heat value of a log before and after it is burned. Similarly, the difference in heat value of petroleum before and after it is burned is a gradient. Energy gradients can be steep – with a great difference from before use to after – or shallow. For the fungus-farming ants discussed in Chaps. 5 and 6, the difference in nitrogen content of insect excrement before and after use is greater than that of leaves. The former has, for ants, a steep gradient. As another example, money is potentially the start of a steep gradient if it is burning a hole in your pocket.

Gradients represent differences in potential energy. Nature tends to diminish gradients, and will dissipate steep gradients with particular efficiency.

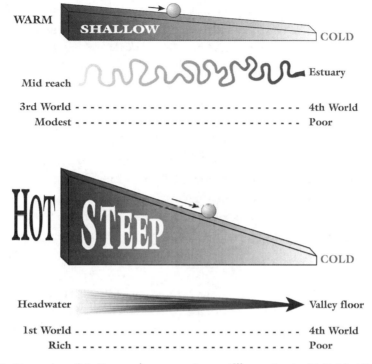

Fig. 9.1 Examples of shallow and steep gradients. (Illustration by T. F. H. Allen; used by permission)

Mountains erode and forests burn. This is in accord with the Second Law of Thermodynamics. To dissipate a gradient is to decrease the magnitude of difference between its starting and ending states, or between adjacent systems with different levels of energy concentration. Place something like the Roman Empire next to some tribal societies, for example, and the members of those tribes will try to dissipate that gradient. This is done through raiding parties whose goal is to loot. Place a country like the United States next to a country like Mexico, and the people of Mexico will work to dissipate that gradient by crossing the border. Place a source of high energy like petroleum in the hands of humans who know how to use it, and we will, quite happily, go about dissipating it. We will continue to dissipate it until the gradient is gone, which is to say until that fuel can no longer be obtained at an energy profit. As described in Chap. 3, we are all merely obeying the Second Law of Thermodynamics.

The steep gradients represented by fossil fuels are part of our cultural context. We grew up with these gradients, we are accustomed to them, and so most people today consider them normal and natural. Petroleum has almost always been cheap in your lifetime and in that of your parents. The energy gradients of fossil fuels are like water or air, a determinant force in our lives that ordinarily we do not perceive. Asking people if their lives are determined by fossil fuel gradients is like asking a fish if its nose is wet. If an energy gradient comes to our attention it is as a Black Swan, a sudden crisis, such as when oil supplies were interrupted during the oil crises of the 1970s, or when prices spiked in 2008. At least, this is the appearance. In fact, the onset of a sudden energy crisis is more like a boiled frog. The processes building up to an energy crisis have been growing in the background for decades, out of sight of most consumers. Then a tipping point is reached, a catastrophe in René Thom's mathematics, and suddenly the world has changed. Similarly, the complexity and riskiness of drilling in open water have been growing for decades, but growing in the background, away from most peoples' sights. So the Gulf spill appeared as a Black Swan when in fact it was a frog finally boiled to death.

Any energy crisis — a price spike, gas lines, or a major spill — comes as a shock because we are socialized not to be aware of the gradients that we are dissipating. Humans and human ancestors lived for perhaps six million years with solar energy as the basis for all we did, which amounts to less than 0.3 watts per square meter. We once knew that energy was hard to get, scarce to humans, and therefore precious. Within those 6 million years, the last 200 years of reliance on fossil fuels are like a grain of sand on a beach or a single star in the galaxy. In other words, the way we live now is an aberration of history, a radical departure from the normal conditions of human existence. Yet we became accustomed to fossil fuels so quickly that now we consider them a normal part of life, an attitude we learn through socialization. We came to assume that energy is inherently inexpensive and naturally easy to come by, that it will always be a steep gradient, and that this gradient will not harm us. Now, we hope, we have learned from the Gulf spill that this gradient *can* harm us. In time, and perhaps sooner than many realize, we will learn not only that it can be dissipated, used up in ordinary language, but that much of it has been.

There will be more energy shocks in the future, shocks of price, supply, and damage. For a while these will continue to be seen as Black Swans. Once they become common enough, as they will, these shocks will become the new normal. At this point, the world in which children grow up will be different, and they will internalize different expectations about energy. Many of our parents and grandparents who grew up in the Great Depression of the 1930s internalized frugality, and lived the rest of their lives that way even if later they became prosperous. Many of us who lived through the oil shocks of the

1970s have had petroleum on our minds ever since. What young people experience matters a great deal. Energy shocks will someday teach our children some harsh lessons. We could ease the harshness of those future lessons if we started now to teach children that high-gain energy (energy that is easy to get and gives a high return) is rare and precious, and that low-gain energy is hard to get in quantity and hard to live on.

Educating our children truthfully about energy may be one of the most important things we can do. In particular we need to teach children to think long-term about energy (something humans don't ordinarily do), to understand the value of history, and to learn to connect the past to the present and the future.

In the remainder of this chapter we draw together the strands of this book, then use what we have learned in order to look toward the future. There is much to ponder. Many problems will converge in our future, problems that will take energy (and the wealth that comes from energy) to address. We will need to develop sources of energy to replace petroleum, or accept living as our ancestors did 200 or more years ago. For that future, we must explore whether our society can survive such a transition in an acceptable form. Most fundamentally, we must bring energy into the public consciousness as something to be aware of always, not just when there is a crisis.

We make no prescriptions here and offer no simple solutions. The reasons should be clear from the discussions to this point: There are no simple solutions. Our purpose, rather, is to prompt a discussion, to help readers to become aware of energy and begin to contemplate and discuss how we will use it in the future.

Why Petroleum Production and Our Lives Are Both Complex

We discussed energy and the Second Law in Chap. 3, and complexity in Chap. 5, including how and why complexity comes about and what it costs. In Chap. 6 we exposed the consequences of complexity, which may be different in the near- and long-term. Complexity and energy gradients interact to exert a powerful influence on major events, including the fates of empires, nations, and civilizations.

Complex systems, as we use the term here, have two parts: these are differentiation in structure (more parts, and especially, more kinds of parts) and organization (limitations on how those parts may behave). Every system is characterized by a degree of complexity, and this complexity can change over time. The late Southwestern archaeologist Emil Haury once asked a group of first-year graduate students at the University of Arizona what they were interested in. When one student answered, "Complex societies," Haury asked, "Do you know of any simple ones?" Point taken: All human societies are complex, but some are more complex than others and nearly all societies today are more complex than they used to be. Why is this? We know that complexity always costs, and the most fundamental cost is energy. In the human past, increasing the complexity of a society meant that people worked harder. The cost of complexity should, one would think, inhibit its development. Yet our societies grow ever more complex. So again the question: Why?

Complexity increases for two primary reasons. When humans have surplus energy, complexity increases because cheap energy allows it. With great quantities of high-quality, inexpensive energy (a steep gradient) people indulge in things that otherwise they could not afford. They travel, eat more, eat more expensively, dress better, and buy electronic widgets that were unheard of a few years ago. Complexity increases, but this type of complexity in particular brings with it the stresses that we know as the complexity of modern life. In the long span of human history, this type of complexification has been rare. Constrained by the limits of solar energy, humans have rarely had surplus energy. When we have had small surpluses, they have been quickly depleted by increases in consumption. Present us with a steep energy gradient and we will dissipate it.

As surplus energy is rare and transient, most of the time complexity increases to solve problems. More sophisticated technologies, more institutions, more kinds of occupations, and more kinds of information—these are kinds of complexifications that can grow in response to problems. Is there a problem of financial regulation? The solution may be new government agencies and new kinds of regulations. Is there a problem of potential terrorism? The answer is new kinds of intelligence gathering, new kinds of security measures, and new screening technology at airports. Are there problems of air quality and petroleum supplies? The response is to develop "hybrid" automobiles with two kinds of engines, which are surely more complex than cars with only one.

Complexity that grows to solve problems typically does so before there is energy to support it. After all, there is a problem and the problem must be addressed. Complexity that grows to solve problems compels increases in energy production. We saw this in the case of Renaissance warfare, as discussed in Chap. 6. Increasing complexity in making war compelled European nations to search for new sources of wealth (i.e., transformed solar energy) around the globe. Increasing complexity to solve problems causes energy consumption to increase. We call this the energy–complexity spiral: increasing complexity causes energy consumption to grow, which in turn brings on higher consumption and new problems, requiring still more complexity and energy (Fig. 5.10).

The fact that complexity and energy consumption grow to solve problems has an implication for modern societies that is of the highest importance. It is that *societies cannot reduce their consumption of energy voluntarily over the long term*. This is a conclusion that will disappoint those who believe that we can secure our future by voluntarily simplifying and consuming less. Although we sympathize with such idealism, to adopt such an ideal would be to assume that the future will present no challenges. Clearly this would be a foolish assumption. As illustrated in Chap. 6, the Byzantine Empire may be history's only example of a complex society surviving through simplification. The Byzantines did not simplify voluntarily; they did it under extreme duress. The energy–complexity spiral suggests that our future will consist of greater complexity and energy consumption, provided that we can find the energy to sustain this. Conventional thinkers believe that innovation can secure our future by reducing forever the energy intensity of the things we consume. In fact, innovation alone cannot secure our future in the long term, as we saw in Chap. 5. Our societies and lives will grow still more complex unless we are forced involuntarily to simplify.

The history of petroleum exploration and production is a history of increasing complexity, increasing costs, increasing risks, and declining net energy. A continual need to solve problems of finding and producing oil has driven this process. In 1892, Edward Doheny, an unsuccessful prospector, found oil near Los Angeles at a depth of 140 meters, close to present-day Dodger Stadium. He drilled the hole using the sharpened end of a eucalyptus tree.[2] From such beginnings, today's drilling technology has evolved. We have discussed this process in Chap. 4 in the specific context of drilling in the Gulf of Mexico. We cannot describe here the entire history of petroleum production, but it is a history like that of other technologies. Problems have arisen, and they have been met by increasing complexity. The biggest problem of any extractive industry is depletion of its most accessible reserves. In the oil industry this means drilling deeper and in more remote places. That first well in Titusville, Pennsylvania hit oil at a depth of 21 meters (70 feet). By the 1920s, wells were being drilled to 2,500 meters (8,200 feet). Today wells

---

[2] http://www.priweb.org/ed/pgws/history/signal_hill/signal_hill.html.

can be drilled as deep as 11,000 meters (35,000 feet). Below that depth the heat of the earth is so high that any hydrocarbons have been cooked away. Another problem is being unable to situate drilling rigs in a desirable place or in enough places, perhaps because of environmental concerns or (especially at sea) because of cost. The solution is horizontal drilling. Horizontal holes can now be bored to distances of nearly 13 kilometers (8 miles). All of this comes at a price, of course. While today's rigs can drill up to 80% faster than those of the 1920s, the scale of the effort has grown exponentially. In 2007, drilling expenditures in the United States alone came to $226 billion, according to the American Petroleum Institute.

The Deepwater Horizon is one of the latest manifestations of the evolutionary process of complexification. Problems such as the depletion of easy deposits and environmental concerns have been met by complexification: the development of technology that is increasingly capable, yet costly and risky. The cost comes not only in the money needed to design, purchase, and run such a rig, but also in the money to repair the environmental damage caused by the Deepwater Horizon spill, and in the damage to a way of life among the Gulf's residents. This is why we refer to proximate (flawed decision making) and ultimate (declining Energy Returns on Energy Invested [EROEI] and increasing complexity) causes of the blowout. Yet despite these costs, we will continue to operate such rigs until they reach the point of economic infeasibility or, more important, the point where the energy returned on energy invested, and the resulting energy and financial balance sheets, make further exploration pointless.

Circumventing the Energy–Complexity Spiral (or not)

The energy–complexity spiral is a system of positive feedback. Each part boosts the other. Energy causes complexity to grow, and higher complexity creates a need for more energy. Although the energy–complexity spiral is not usually recognized explicitly, many writers about our future recognize it implicitly. And as we discussed in Chap. 5, people understand complexity because we experience it in our daily lives.

Can anything be done about the energy–complexity spiral without diminishing our material quality of life? Two things commonly suggested are conservation and innovation. In fact these are related, for one focus of innovation is to improve the efficiency of our technology by designing devices that use less energy to achieve the same output. But does either conservation or innovation provide a way out of the energy—complexity spiral?

As discussed in Chap. 5, when people experience a reduction in the prices of things they like to consume, the ordinary response is to consume even more. William Stanley Jevons pointed out in the middle of the nineteenth century that improving the efficiency of a technology does not promote conservation of fuel. To the contrary, efficiency causes prices to fall so much that we consume more of the fuel than ever before. Over the long run, efficiency causes more of a fuel to be consumed than would otherwise be the case. Tainter likes to challenge his students with a Jevons-type paradox. Suppose, the students are asked, that you are going to buy a car. The choices are a fuel-efficient hybrid or an enormous sport-utility vehicle. Which do you choose? A few students will honestly say that they would like to have the SUV, but most pick the hybrid. The latter group are conscientious, or at least want their peers and professor to think so. But here is the paradox. If everyone in America drove an SUV we might consume less fuel than if everyone drove a hybrid. Those SUVs would be parked much of the time whereas the hybrids, being economical, would get a lot of use.

Writing for *Macleans*,[3] Andrew Potter gave us an article cunningly titled, "Planet-friendly design? Bah, humbug." The story's lead reads, "The chief result of energy-efficient housing is in fact the rise of McMansions." As manufacturers have improved insulation and the efficiency of furnaces, people have not responded by buying energy-efficient houses. Instead, in the first decade of the twenty-first century, people just bought larger houses, giving rise to the phenomenon of the McMansion. "Our consumption habits," writes Potter, "seem to be ruled by a principle of 'waste homeostasis,' where the energy savings we get from better technology is used to fund better toys." Sustainability, suggests Potter, "…is not a matter of how things are designed, but of how they are used."

A remarkable doctoral dissertation written by Eva Andersson in 2000 at Umeå University in Sweden attempted to model what would happen if Swedish households were to adopt a "green" lifestyle: less meat consumed, less use of private cars, efficient housing, and the like. This would all save money. What would Swedes do with the money thus saved? Andersson's study suggests that if Swedes lived such a lifestyle, they would like to use the

---

[3] http://www.macleans.ca/article.jsp?content=20070202_154815_4816.

money to travel more. Because travel involves fossil fuels and carbon dioxide emissions, green living in this case would simply shift energy consumption and environmental impact from one sector to another. William Stanley Jevons would have predicted it. Andrew Potter might call it waste homeostasis.

Does this mean that conservation is not worthwhile? Of course not, nor do we wish to discourage it. Conservation helps, but it does so in the short term. The trick to circumventing the Jevons paradox and making conservation work in the long term is to decouple conservation from saving money. If conservation were to be necessary but people could not save money through conservation, the Jevons paradox and waste homeostasis would be negated. Astute readers will point out that this of course eliminates much of the incentive to conserve. That is true, and it is the first of many points on which we must begin to have serious discussions at the national level.

There is always innovation. Technological optimists believe that innovation will offset future resource shortages by wringing ever more efficiency from our technologies. Hybrid automobiles are one example of this, and surely there are many more innovations to come. As with conservation, though, it appears that innovation will not be effective in the long run. As discussed in Chaps. 3 and 5, the productivity of our system of innovation has declined by over 20% in only one generation (Fig. 5.5). Innovation in the energy sector parallels the declining productivity of innovation overall (Fig. 5.7). This includes the wind and solar technologies that we will presumably need in the future. None of this is to say that innovation will soon disappear. But if present trends continue, within one or two generations our system of innovation may become too expensive and too unproductive to continue. Just as we have depleted the most accessible pools of petroleum, we have also depleted much of the stock of intellectual breakthroughs that are easy to make. Penicillin no longer waits to be discovered, and the internal combustion engine no longer waits to be designed. Even new sectors such as biotechnology and nanotechnology are showing declining productivity of innovation (Fig. 5.6).

Both conservation and innovation have implications that will be different in the short term and the long term. One lesson is that, in considering our energy future, we must force ourselves to think long term. By that we mean in terms of decades to generations. Thinking long term is something that humans are not normally inclined to do. We will need to overcome this disinclination.

EROEI: Background for Thinking About the Future

We now understand that it takes energy to get energy. The ratio of what one gets to what one expends is referred to as energy returned on energy invested. Where the net energy is great, one can think of this as a steep energy gradient (Fig. 9.1). Steep gradients easily accomplish work: Think of rolling a car downhill. Among other things, a steep energy gradient can power a complex society quite easily. A shallow energy gradient can power a complex society too, but not as easily. Because the net energy is low, it takes organization to enable consumption (dissipation) of the same amount of useful energy, as well as a greater amount of energy overall. In other words, running a complex society on a shallow energy gradient will be more energy intensive than running one on fossil fuels. This fact is never discussed in the mass media. This is the lession of the Roman Empire when it came to rely on yearly taxes rather than conquest (Chap. 6), and of the leaf-cutter ants who need highly organized societies to transport enormous numbers of leaf fragments (Chap. 5).

Most sources of renewable energy have less energy density, and a lower EROEI, than do fossil fuels. Relying on low-gain energy means using more and more of the earth's surface for energy production. The best recent data on renewables have been compiled by David MacKay in his book *Renewable Energy – Without the Hot Air*. The average American uses 250 kilowatt hours per person per day (kWh/p/day) from all energy sources. Most of this, as we know, comes from fossil fuels. If 425,000 square kilometers of our windiest places (North Dakota, Wyoming, and Montana) could be converted to wind energy production, this would produce 42 kilowatt hours per person per day if allocated among 300 million people. Offshore wind production taking up a large part of the shallow waters of the U.S. East Coast could produce another 4.8 kWh/p/day. Geothermal energy might give us 8 kWh/p/day (but see below for an associated risk). The combined hydroelectric facilities of Canada, the United States, and Mexico can yield 7.2 kWh/persons/day. Adding these up we get only 62 kWh/p/day, less than one-fourth of what we need. Moreover, achieving even this amount of energy production by wind would require us to fill more than half the area of Texas with wind turbines. This cannot be achieved in practice.

Don't count on biofuels either. Consider that most biomass accumulated each year on the earth by plant photosynthesis is used as food by all nonphotosynthesizing inhabitants of every large ecosystem on our planet, all animals (humans included), insects, fish, fungi, and bacteria. Using NASA's satellite estimates of global photosynthesis, Patzek has shown that our Earth does not have much spare biomass production capacity, and humans have already intercepted more than she can offer without being injured. In particular, the Earth has no capacity to produce giant quantitites of biofuels that could rival fossil fuels in volume every year for many decades to come.

What about solar energy? Either photovoltaics (solar panels) or solar concentrators (which heat oil in pipes) literally fill the land surface on which they are installed. An area of 350,000 square kilometers, devoted to solar energy production in the sunniest parts of the U.S., could deliver electricity equal to the current continuous (24 hours per day) U.S. consumption of 250 kWh/p/d. But such gross figures are misleading. Firstly, of course, it is dark half the time. Secondly, the effectiveness of a solar plant increases from zero at dawn, to four to five times the average power at noon, and back to zero at dusk. Thirdly, it is not possible to transmit all that power from a few parts of the Southwest to the rest of the U.S. But the biggest problem is the challenge of building and maintaining this infrastructure. Consider the solar plant shown in Fig. 9.2. It is the sort of plant that many consider our hope for the future, and also a large footprint on the landscape. Yet, according to Patzek's calculations,[4] we would need 211,000 such plants to power the U.S. by solar power alone. Imagine the job of cleaning dust off the collectors, if we could even build so many. At the same time, we would need conventional generating plants fueled by coal, natural gas, or uranium to supply electricity at night.

The United States is actually well endowed to produce renewable energy compared to other places. MacKay shows that for England, Scotland, and Wales to produce all the energy renewably that they currently need would require most of the land of England, Scotland, and Wales.

Renewable energy also means that there need to be complex systems of organization to aggregate and store small amounts of net energy from each production source. Some of this organization would come through physical infrastructure such as electrical cables that in turn require massive quantities of fossil energy to be produced and deployed. Some of it would also come through "soft" infrastructure: public or private bureaucracies to ensure collection and distribution, and manage storage and load. Both hard and soft infrastructure will have energy costs.

---

[4] Patzek, Tadeusz W. Earth, Energy, Environment and Economics. Second Edition. Manuscript in preparation, University of Texas, Austin.

Fig. 9.2 The Nevada Solar One plant, a solar energy concentrator that covers 1.6 square kilometers of land and provides 15 megawatts of electricity on average. Its vastness can only be appreciated from the air. The figure of 15 megawatts is obtained by dividing the number of kilowatt-hours actually sold in one year by the number of hours in a year. At today's power requirements, this plant can serve 1,440 people on average. We would need to build 211,000 such plants to generate 250 kWh/p/day, which is current U.S. consumption

Renewable energy production will damage the earth. We should not ignore this fact. There is no way to avoid it. Being extensive in its land requirements, low-gain energy is always destructive. Although we have greatly damaged the earth with our use of fossil fuels, by far the greatest alteration of the earth's ecosystems has come from low-gain agriculture. It takes a lot of the earth's surface to power a large complex society from shallow-gradient energy. Similarly, ants that strip leaves from trees are often considered an agricultural pest, interfering with human cultivation. Production of energy from low-gain renewables will not only consume much of the earth's surface, it will also cause much conflict over land uses.

Bearing these points in mind, it is clearly important that the earth's people begin an adult discussion about EROEI, future energy, and future complexity. We say an "adult" discussion to emphasize that this should be a realistic discussion devoid of the usual punditry, and rising above the self-serving canards uttered by so many of our politicians. The matter is too important to accept anything less. A future collapse due to insufficient energy is too gruesome to contemplate. In the next few pages we offer some options and alternative scenarios. As noted previously, there are no easy answers and we offer none. We hope what we do offer is grist for discussion of energy policy.

Energy and Complexity in Our Future

We are not the first people to face an energy dilemma. In Chap. 6 we reviewed three examples of societies that faced problems of energy and complexity. Each found different solutions to their problems, and from this experiment we can foresee possible options for ourselves. Let's review them briefly.

Responding to near-fatal crises in the third century A.D., the Roman Empire increased the size and complexity of its main problem-solving institutions, the government and the army. The complexification worked, as complexity often does, and the empire survived for another two centuries. But complexity always costs, and however we account for it, the ultimate cost is energy. Being dependent on solar energy, there was little the Romans could do to promote growth in their economy. Instead, the government established production goals for every plot of land across the empire, and raised taxes to levels that subsistence farmers could not pay on a continuing basis. The result was to undercut their energy production system. Farmers who could not pay their taxes abandoned their lands and went to work for wealthy landowners. When the government required town councils to ensure that taxes were paid, elites left cities and towns to live in their rural villas. The Roman Empire went from living on the equivalent of yearly interest – the yearly production of the empire's agricultural lands – to consuming its capital resources: productive lands and peasant population. The ultimate collapse was inevitable.

Part of the Roman Empire did manage to carry on in the eastern Mediterranean, and we know it today as the Byzantine Empire. Faced in the seventh century A.D. with the loss of half their lands to the Arabs (i.e., the loss of half of their energy base), the Byzantines responded with one of history's rarest innovations: they systematically simplified their government and army, greatly reducing the energy needed to run their society and empire. The simplification worked. The Byzantine Empire began to recover, and in time recaptured some of the lands that had been lost. An energy–complexity spiral can trend both upward and downward. When it trends downward with less energy, necessitating less complexity, requiring less energy, the experience is usually unpleasant. The Byzantines alone made it work to their advantage.

Warfare in Europe over the past few centuries caused energy and complexity to spiral upward. Competition forced European states continually to innovate in armaments, tactics, organization, and logistics; to increase greatly the sizes of their armed forces; and to allocate more and more of their wealth to war. This was done on the basis of an economy that, as did the Roman Empire, ran until recently on solar energy. Like the Romans, European states taxed the peasants, so heavily in fact that popular revolutions in the eighteenth and nineteenth centuries led to the democratic societies of Europe today. How did the Europeans avoid the fate of the Romans? The answer is that they found subsidies to pay the cost of their energy–complexity spiral. Over the horizon they found whole new continents to conquer. As much as possible, the wealth of those continents (transformed solar energy) was transferred to Europe. In 1622, to give just one example, the Spanish ship *Nuestra Señora de Atocha* and five other ships sank in a hurricane in the Florida keys. The *Atocha* alone carried over 900 silver bars, weighing 35 tons, and 161 pieces of gold. This was valued at the then-astronomical sum of one million pesos. When discovered in 1985 the treasure proved to be worth $450 million. This type of subsidy, tapping energy earned by others, is how Spain financed its European wars. Yet as the Romans found, such booty can finance a complex society only so long. Eventually looted treasure (like pools of petroleum) runs out. More recently, all societies of today, led by Europe, made the transition to financing themselves through fossil fuels, supplemented to varying degrees by nuclear power and a few other sources. This continues the European tradition of financing complexity through subsidies—energy coming from elsewhere. In this case, the "elsewhere" is the geological past, and mainly of regions other than Europe itself. The energy–complexity spiral is most powerful when its requirements are subsidized by cheap energy that some other people, or place, or time has accumulated for us.

Which of these strategies are modern societies likely to pursue in the future? The Roman model of robbing Peter to pay Paul leads, in the end, to fiscal weakness, disaffection, and collapse. It can be followed for only a short time, and we must hope that we never need to adopt it. The Byzantine model of simplification and conservation is a strategy that humans seem willing to adopt only when there is no alternative. It may lie someday in our future. To be sure, we will try to continue the European model of energy subsidies for as long as we can. Humanity will not forgo such rich, steep gradients. Even the threat of climate change will not deflect humanity from searching for oil in ever-more-inaccessible places, nor from burning through our mountains of sulfurous coal. Too many people find the short-term wealth and well-being irresistible. Too many of us are fish immersed in water that we do not perceive, the "water" being the steep energy gradients to which we are accustomed. In making this observation, we are not approving such behavior. We are merely predicting (realistically, we think) that humanity will behave in the future as we have in the past. We will use fossil fuels as long as we find it economically and energetically feasible to do so. For how long, though, can we follow the European model? Declining EROEI suggests that the answer is: Not forever.

Herbert Stein, a noted economist and chairman of the Council of Economic Advisors under presidents Nixon and Ford, once said, "If something cannot go on forever, it will stop." Known as Stein's Law, this is sometimes expressed as, "Trends that can't continue, won't." Our use of fossil fuels cannot continue forever, and it won't. As we search for petroleum in some of the most remote places on earth, for which we need technology that is expensive and risky, the EROEI (net energy) declines. In the United States in the 1940s, petroleum exploration and production gave an EROEI of 100 to 1. Today, as assessed by Charles Hall and his colleagues, it averages about 15 to 1 (see the Further Reading list at the end of the chapter). Already for oil from the Canadian tar sands the net energy can be as low as 1.5 to 1. It takes the energy of about two barrels of oil to produce three barrels of conventional oil from tar sands. In a pioneering analysis, Charles Hall and his colleagues estimated that we need net energy of at least 5 to 1 to power a modern complex society (see Further Reading). Fortunately, not all petroleum sources have an EROEI as low as that of tar sands. That is, however, the direction in which we are headed. Someday, the physics of net energy will curtail our use of petroleum. A trend that cannot continue, won't. And as we have discussed, it appears that innovation may not offset declining net energy to the degree that we would need it to over the long run.

Yet while net energy declines, the problems that our societies face demand resources and increasing complexity, complexity that takes energy. In the next few decades, the United States and other industrialized nations will confront a convergence of very expensive problems, problems that will require wealth and energy to address. We discuss here seven such challenges, acknowledging that there may be others of equal magnitude. These challenges are: (1) funding retirements for the baby-boom generation; (2) continuing increases in the cost of health care; (3) replacing decaying infrastructure; (4) adapting to climate change and repairing environmental damage; (5) developing new sources of energy; (6) in all likelihood, continuing high military costs; and (7) innovation.

Retirement Costs: The annual cost of Social Security benefits came to 4.8% of the United States' gross domestic product (GDP) in 2009. The cost is expected to reach 6.1% of GDP in 2035. In 2010 the system paid out more than it took in, and its surplus is projected to be exhausted in 2034. The primary reason, as is well known, is that 80 million baby-boomers are now beginning to retire, and they are living longer than retirees used to do. There once were 16 workers who paid Social Security taxes for every retiree. When the last boomer retires, that ratio is expected to be 2 to 1. We will need cheap abundant energy to meet these obligations. Social Security is not the only problem. Many current workers have invested in the stock market to fund their retirement. In the financial crisis of 2008–2009, much of that investment disappeared.

Health Care Costs: Similarly, the cost of health care will rise with the aging of the baby-boomers, and with the advance of medical technology. The medical products industry continually brings out wonderful new things, products that restore health and save lives. Yet it is precisely these marvelous, energy-intensive technologies that are driving the cost of health care so far ahead of the economy as a whole. Health care costs have risen from 7.2% of GDP in 1970 to 16% in 2005. These costs are projected to rise to 20% of GDP in 2015. Will the cost of health care in 2015 also require 20% of our energy production?

Replacing Infrastructure: In 2005 the American Society of Civil Engineers gave American infrastructure (roads, buildings, pipes, airports, schools, and the like) a grade of D, down from D + in 2001. Much of our infrastructure is aged and in need of refitting. This problem was brought to the nation's attention on August 1, 2007, when the Interstate 35 bridge across the Mississippi River in Minneapolis suddenly collapsed, killing 13 people (Fig. 9.3). Some 700 U.S. bridges are of similar construction. Our current expenditures on existing infrastructure reflect an approach sometimes called "patch and pray." The engineers estimate that we will need $1.6 trillion to restore the country's infrastructure to safe and proper conditions. Someday we will have to pay this, or do without the roads, bridges, and schools that we are accustomed to using. No one has calculated the energy cost of either option.

Climate Change and Environmental Damage: What will it cost the world in money and energy to resettle tens of millions of people who will be flooded out of coastal Bangladesh if projections of global warming are correct? This is only one of multiple costs that warming will impose upon the world. Will we need to build a seawall around Manhattan Island? Looking only at four cost sectors – hurricane damage, real estate losses, energy sector costs, and water costs – Frank

Fig. 9.3 Collapsed Interstate 35 Bridge in Minneapolis, August 31, 2007. (Source: Wikimedia)

Ackerman and Elizabeth Stanton of Tufts University estimate that global warming may cost the United States $271 billion in 2025, rising to nearly $1.9 trillion in 2100. This will be an estimated 1.8% of U.S. GDP, equivalent to a recession, but a mild one. Estimates get worse from there. Total world economic damages, according to some researchers, could total 6–9% of global economic output, equivalent to a recession of the magnitude of that of 2008–2009. (That global economic output is, of course, transformed energy). Yet according to Claudia Kemfer and Katja Schumacher of the German Institute of Economic Research, damage could actually range as high as 20% of world economic output, which is in the range of the 1930s Great Depression. Added to this will be whatever we need or choose to spend on repairing the damage to our air, soil, and water from past economic activity and energy production. Just to clean up the poly-chlorinated biphenyls (PCBs) dumped by General Electric Company into the Hudson River is estimated to cost $500–600 million, and that is just one prob-lem in one river.

Developing Renewable Energy: Renewable energy is in one way a little like hybrid automobiles. Just as a hybrid car needs two engines, so plants that produce electricity from wind or the sun need back-up plants powered by coal or natural gas for times when the wind doesn't blow or the sun doesn't shine. Low-power renewable energy sources also need the massive subsidy of fossil fuel power to be manufactured and deployed. To convert modern societies to renewable energy will be a costly enterprise, a problem exacerbated by the low EROEI of most alternative sources of energy. Crude oil, once giving an EROEI from large onshore reservoirs in the range of 100 to 1, now yields about 18 to 1 globally, and identical oil from the United States averages about 15 to 1. Electricity from the sun, in the most sun-drenched places, can give an EROEI of about 6.5 to 1, and in the airiest places wind generation can give as much as 20 to 1. That last figure sounds good, but it is a figure achievable only in the best locations and at the best times. Overall, renewable sources tend to have low power (watts per unit area), and it takes complex technologies to make the best use of them. Complex technology is expensive: in dollars, in energy, and in environmental damage. A high-tech Tesla roadster (a sports coupe driven by an electric motor) has a lithium-ion battery pack weighing 900 pounds that can deliver 190 megajoules of energy. A conventional automobile with a 10 gallon tank that weighs 62 pounds can deliver 1,200 megajoules of energy. A 1,000 megawatt conventional power plant requires one to four square kilometers of land. According to Patzek's calculations, a solar plant capable of generating the same amount of electricity would need up to 625 square kilometers. Texas's Horse Hollow Wind Energy Center is the largest wind facility in the U.S., with a nominal capacity of 735 megawatts. It is spread across approximately 190 square kilometers of Taylor and Nolan counties, near Abilene. The average turbine power is at best 1/4 of its peak power, ranging down to zero on hot summer days. To generate 1000 megawatts on average with wind turbines requires over 1,000 square kilometers of land. Patzek[5] has calculated the average power of ethanol produced from a switchgrass field in the U.S. To obtain 1,000 megawatts of heat from switchgrass ethanol would require about 8,600 square kilometers of land. If the ethanol were converted to 1,000 megawatts of electricity, the area required for switchgrass would become 25,000 square kilometers. It would make more sense to burn the switchgrass and generate electricity directly. Renewable energy will impose high investment and environmental costs in the next few decades. We will need massive power from fossil fuels to subsidize that transition, rather like a large truck pushing a tiny car.

---

[5] gaia.pge.utexas.edu/papers/SustainabilityTWP092Published.pdf

Military Costs: We explored in Chap. 6 the factors driving increasing complexity and costliness in military technologies. Norman Augustine, aerospace engineer and undersecretary of the Army from 1975–1977, once wrote facetiously of this trend:

> In the year 2054, the entire defense budget will purchase just one aircraft. This aircraft will have to be shared by the Air Force and Navy 3-1/2 days each per week except for leap year, when it will be made available to the Marines for the extra day.

Given the advantage of complexity in military technology, there will be pressure for this trend to continue. Unless the peoples and nations of the earth become uncharacteristically amiable and selfless in the next 20 or 30 years, the only thing that might disrupt this trend is poverty or lack of energy. Both the American and German militaries have already considered the implications of peak oil, the point when the rate of oil production inevitably decreases.

Innovation: To overcome the myriad problems that we face we will need to continue to invest in research and development. As shown in Chaps. 3 and 5, however, the productivity of innovation is declining. This means that we must invest more and more to get a constant output of research products. Responding to this problem, a few years ago Congress doubled the budget of the National Institutes of Health, and now proposes to double the budget of the National Science Foundation. As the productivity of these investments declines, we will need to continue to spend increasing amounts, or accept fewer innovations. Derek de Solla Price anticipated Norman Augustine's reasoning when he wrote in 1960 that if we continue to allocate more and more of our resources to research, the day will come when we must all be scientists. The national wealth that allows us to invest so much in research and development comes, of course, from energy use.

Each of the above is an expensive problem that, if occurring in isolation, we could probably afford to solve. The challenge is that they will converge almost simultaneously over the next few decades, in a "perfect storm" like nothing the movie industry has imagined. These problems in combination will be extremely costly if they must all be addressed at once, as it appears they will. We see three major threats coming from this challenge.

The first challenge is that an economy is a pie, as is the energy budget on which it is based. From this pie, shares are allocated to different sectors of a society: consumption, investment, infrastructure, innovation, government, and so on. Increasing the share of the pie going to one sector will mean proportionately less for all others. If we must invest a larger share of our wealth (read: energy) in, say, health care or innovation, there will be correspondingly less to spend on something that matters to people: consumption, which economists frequently argue drove a large part of our twentieth and early twenty-first century growth economy and the jobs it supports. The cost of addressing the seven problems may be a proportionate decline in our material quality of life. People will not accept this quietly. Conventional thinkers will argue that the solution is to increase the size of the pie, so everyone is satisfied. This is fine, if somehow it can be brought about despite declining EROEI and diminishing returns to innovation, and without further environmental damage.

The second threat is the same as the Romans faced in the third and fourth centuries A.D.: increasing complexity and costliness just to maintain the status quo. The seven problems are challenges that we must face mainly to keep our lives as they are. The one possible exception is innovation, which for some time will have the potential to continue to improve our lives. The others (retirements, health care, replacing infrastructure, adapting to climate change, developing new energy, and military costs) are expenditures that we must undertake to keep our lives as they are. As the Red Queen said to Alice in *Through the Looking Glass*, "Here, you see, it takes all the running you can do, to keep in the same place." Paying more and more to maintain the status quo is the very essence of diminishing returns to problem solving. The Romans found this to be a strategy that in time produces fiscal weakness and disaffection of the population, if undertaken with an energy base that cannot grow.

The final threat is another way of looking at the first two. It is that money is transformed energy. We look at the cost of solving these converging problems as a financial challenge, but we know from Chaps. 2–5 that it is actually a challenge of energy. Energy generates the wealth to solve problems, these seven or any others that may come up.

Here is the big challenge, the mother of all problems: these converging problems will themselves converge with other physical and societal trends. Problems, we now understand, generate increasing complexity and costs. To address the seven problems will require not just higher expenditures, but also greater shares of our energy and financial resources. Yet these converging problems come upon us just when our investments in petroleum production are themselves producing diminishing returns. We must now look for oil in places that are deep, dark, cold, remote, and exceedingly risky. It takes a lot of energy to find and produce energy in such places. Just when we need more energy to address the seven problems, our net energy, EROEI, is declining. In other words, we may have less capacity to generate wealth just when we will need it most.

Declining net energy and declining productivity of innovation lead us to wonder if our future will be a steady-state economy, as Herman Daly has envisioned it, at least for a time. We know from the work of Cutler Cleveland,

Charles Hall, James Brown, and their associates that energy per person is the fundamental ingredient of economic growth. It is not clear that energy per person can rise in the future as it has risen over the past two centuries. There are reasons to think that it may not or, as we discuss below, may not easily. Innovation is how we have turned energy into wealth. Many people count on innovation to offset declining resources per person. But if energy per person will not increase, and if innovation continues to become less productive, the engines of economic growth will disappear. The result would then be an economy that does not grow, a steady-state economy.

Daly defines a steady-state economy as "...an economy with constant stocks of people and artifacts, maintained at some desired, sufficient levels by low rates of maintenance 'throughput,' that is, by the lowest feasible flows of matter and energy from the first stage of production to the last stage of consumption." Steady-state means exactly that: Consumption is flat. Employment is level. Throughputs of matter and energy are fixed. Birth rates equal death rates. Savings and investment precisely equal depreciation. Complexity does not increase. There is no energy–complexity spiral. This is also sometimes known as full-world economics. In a steady-state economy, dreams of personal advancement would consist once again of such serendipity as somehow marrying a princess. A steady-state economy has been anticipated by philosophers since the eighteenth century, and explicitly formulated in the nineteenth century by John Stuart Mill: "The end of growth leads to a stationary state." Unfortunately, as pointed out in Chap. 3, even a steady-state economy will in time become a gradually collapsing economy, because of the accumulation of chemical waste on the earth and the merciless Second Law of Thermodynamics. That is, a so-called steady-state economy would be steady only for a time. A steady-state economy is impossible to maintain over the long-term. In time, a steady-state economy would become a steadily shrinking economy.

We are not advocating a steady-state economy, merely noting that the laws of physics and the complexity of innovation may force one upon us. In fact, we have further reservations about such an economy. One reason is complexity in problem solving. Human societies, as we have emphasized throughout this book, increase in complexity to solve problems. This may be in the realm of technology, in social or political structure, in the production of goods and services, or in the production and distribution of information. Complexity consists of generating new kinds of parts to a system and new kinds of organization, and it always has an energy cost. Societies increase in complexity to solve problems, and subsequently must produce more energy and other resources to pay for the increased complexity. In a steady-state economy this would not be possible. A steady-state economy is the antithesis of how humans solve problems. Energy and other resources in such an economy are constant and cannot be increased. Complexity in any sector therefore cannot grow except by taking resources from elsewhere within the society, robbing Peter to pay Paul. This is what the Romans did, and it caused them to shift from living off interest – yearly agricultural production – to consuming their capital resources: productive lands and peasant population.

To advocate a steady-state economy is to assume that the future will hold no challenges and all waste will be perfectly recycled. If this proves unfounded, as undoubtedly it will, we may find that we cannot solve our problems no matter how pressing they may be.

It remains to say a few words about a future based on a so-called "green economy," an economy powered by energy sources that are renewable and, perhaps, carbon-free. It is fashionable in some quarters to see such an economy as the key to a prosperous future, a fountain of new industries and new kinds of jobs. We point out that this scenario, like any route to economic growth, requires either innovation (which we know is diminishing) or increasing energy per person. Without increasing energy per person and with less innovation we would be in a steady-state economy, with the challenges just discussed. As noted above, to produce 1,000 megawatts of electricity from renewables, instead of coal or natural gas, requires between one hundred and several thousand times more land, with concomitant environmental damage. Faced with such daunting figures, it is not clear whether renewable energy can produce even a fraction of the power per person that we enjoy now, let alone more energy to solve the problems that we will inevitably confront. Renewable energy will go through the same evolutionary course as fossil fuels. We will first put to use the best sources in the best places. If population increases or we need to increase energy per person, we will next look to sources and places that are less suitable. This will take more land area per unit of energy cap-tured, and probably increasingly complex technology. The marginal return to energy production will decline, just as it has with fossil fuels. Whether renew-able energy can fulfill our expectations is an experiment that will play out in the rest of the twenty-first century.

We stated that we would not offer simple solutions. There aren't any. Neither have we discussed the staggering energy-demand problems brought upon us by the growing human population. What we have offered are sce-narios of alternative futures that we hope will provoke an assessment of our future that is honest and realistic. We owe it to our children and grandchildren to develop that assessment without further delay.

The Deepwater Horizon Blowout: Proximate and Ultimate Causes

No doubt many individuals made decisions on the Deepwater Horizon that they would later regret. There were pressures of time and money. Promotions and careers may have hung in the balance, not to mention profits, stock prices, and dividends. There was probably an intangible element too, the understandable desire of professional workers to get a job done. Fatigue and frustration from a job that had gone on too long may have played a role. Some of these are things that we may never know. But we do know the incentives and pressures to which humans respond, and as the drilling of the BP Macondo well went on and on, no doubt those factors became paramount.

But that is not the end of the story. These are the proximate reasons for the blowout, the factors specific and immediate that caused the well to fail. The hanger option was not implemented, too few centralizers might have been used, the cement was questionable, no cement bond log was run, different muds made pressure readings difficult, and of course the blowout preventer did not fully deploy. What these proximate factors do not tell us is why, to begin with, we are looking for oil in risky places. To understand that larger question we must explore ultimate causes.

Proximate causes tell us that humans are still human: self-serving, short-sighted, and fallible. Ultimate causes are harder to discern, and require us to weave together many disparate strands. In the wide-ranging course of this book, we have searched for these strands. Brought together they tell us much about our society and its direction.

The Principle of Least Effort, basic economics, and growing complexity explain why we look for oil in remote places. Least effort tells us that humans first use the resources that are least costly to acquire and process before using resources that are more so. Economics reinforces this elementary lesson, for people will not ordinarily choose difficult options when easier ones will suffice. Economics also cautions us to pay attention to *marginal* changes in return on investment, changes in the net return when an investment is increased by a specific amount. Complexity comes about on its own, almost unwittingly, as we develop solutions to the problems that confront us.

We know that people develop more complex technologies and institutions to solve problems. Complexity consists of designing systems that have more parts, more kinds of parts, and more organization. Complexity always has costs, and ultimately those costs are paid by energy. The Principle of Least Effort dictates that less complex and costly solutions are adopted first, and more complex solutions are developed when new challenges arise. As we have depleted the least costly reservoirs of petroleum, we have been forced to search for oil in ever more remote places: deeper in the earth, in the Arctic, offshore, and now in deep oceans. As we have looked for petroleum in these realms, we have needed to develop more complex technologies to accomplish this. Unfortunately these technologies tend to be costly and risky. As Jad Mouawad and Barry Meier described recently in *The New York Times*,[6] technologies of extraordinary complexity and capability have been emplaced in the Gulf of Mexico. Mouawad and Meier note that many of these technologies are "far more sophisticated than the ill-fated BP well that blew up in April." Perdido, a rig costing $3 billion, can pump oil from 35 wells across 30 miles of ocean floor at a rate of up to 130,000 barrels a day, while simultaneously drilling new wells. Its deepest well lies in 9,600 feet of water. All the while, the EROEI, net energy, from American oil production has declined from 100 to 1 in the 1940s to 15 to 1 today.

In 1984, an ironic year, Charles Perrow published *Normal Accidents: Living With High Risk Technologies*. Written in an era of simpler technology than today, its lessons have grown more applicable with each passing year. Perrow uses the term "normal accidents" partly as a synonym for "inevitable" accidents, accidents whose likelihood is inherent in a complex technological system. In a highly complex piece of technology with many parts, accidents happen from unpredictable interactions among some of those parts. Complexity makes failures nearly inevitable. Engineers try to avoid failure by adding more complexity, all of which makes the operation of technological systems difficult for human operators to understand. Consider the challenges Toyota has recently had diagnosing the problem of unintended acceleration in some of its vehicles, let alone the fact that its engineers did not anticipate these design problems (and perhaps did not find them in the end). Safety systems may actually create new kinds of accidents, as happened at Chernobyl where faulty tests of a safety system caused the meltdown and fire. Human operators, not understanding what is happening in a complex accident, are vulnerable to taking actions that make things worse.

On March 18, 1967, the oil supertanker *Torrey Canyon* went aground on a reef near the coast of Cornwall, England. It was carrying 119,000 tons of Kuwaiti crude oil. Some 50 miles of French coast and 120 miles of Cornish coast were contaminated, and around 15,000 sea birds killed. The story of how the *Torrey Canyon* went aground is illuminating. According to *The Times Atlas of the Oceans*,

---

[6] Mouawad, Jad and Barry Meier. 2010. "Risk-Taking Rises as Oil Rigs in Gulf Drill Deeper." *The New York Times*, August 29, 2010.

At 08.42 the master switched from automatic steering to manual, and personally altered the course to port to steer 000°T, and then switched back to automatic steering.

At 08.45 the third officer, now under stress, observed a bearing, forgot it, and observed it again. The position now indicated that the *Torrey Canyon* was less than 1 nm from the rocks ahead. The master ordered hard-to-port. The helmsman who had been standing by on the bridge ran to turn it. Nothing happened. He shouted to the master who quickly checked the fuse – it was all right. The master then tried to telephone the engineers to have them check the steering gear aft. A steward answered – wrong number. He tried dialling again – and then noticed that the steering selector was on automatic control instead of manual. He switched it quickly to manual, and the vessel began to turn. Moments later, at 08.50, having only turned about 10°, and while still doing her full speed of 15-3/4 knots, the vessel grounded on Pollard Rock.[7]

This was a normal accident, an accident caused by technology and confusion (exacerbated no doubt by the fact that the ship's cook was at the wheel). The *Torrey Canyon* was technologically simple compared to the Deepwater Horizon.

Complexity was clearly a problem on the Deepwater Horizon. One emergency system was controlled by 30 switches. The rig had multiple safety systems but, as recounted by David Barstow, David Rohde, and Stephanie Saul in *The New York Times*, every one of the safety systems failed. Some did not work, some were activated too late, and some were not activated at all.[8] Crew members hesitated and did not take decisive steps. Communications failed. Warning signs were missed, and crew members in critical areas did not coordinate a response. Some of these failures are clear signs of systems that were too complex to be deployed in a life-threatening situation. As the board that investigated the loss of the space shuttle *Columbia* noted, "Complex systems almost always fail in complex ways."

Normal accidents appear as if they are Black Swans, something that cannot happen. In fact, the very nature of complex technologies makes accidents probable. They are a normal byproduct of the operation of systems whose complexity is beyond human understanding. This observation surely applies to the technological wonders currently operating in the Gulf of Mexico. As described in *The New York Times*, the oil exploration industry has entered an era of greatly increased complexity and concomitantly increased risks.[9]

---

[7] Alaistair Cooper (ed.) 1983. *The Times Atlas of the Oceans*. Van Nostrand Reinhold, New York. p. 169.

[8] David Barstow, David Rohde, and Stephanie Saul, 2010. "Deepwater Horizon's Final Hours." *The New York Times*, December 25. http://www.nytimes.com/2010/12/26/us/26spill.html.

[9] Mouawad and Meier, *op. cit.* (in note 6).

The risks will surely continue to increase. There is talk now of vast deposits of oil off the coasts of Greenland, even to the north of the arctic environment that sank the *Titanic*.

These, then, are the ultimate causes of the Deepwater Horizon blowout. Having depleted easily accessible oil, we must search for oil in places that are increasingly more remote and challenging. To do this we develop technologies that are complex, costly, risky, and difficult to comprehend, in parallel with the overall complexification of our societies. All of this is done in a search for energy that gives lower net returns, so that we use more and more petroleum in our search for the same stuff. We do not want to say that the Deepwater Horizon accident was inevitable. But the system of which it is a part, and the societal pressures driving that system, made such an accident highly likely.

Final Remarks

Looking back at what was done in the final days of work on the BP Macondo Well, we may think that the blowout could have been prevented. But the proximate and ultimate causes of the blowout would still have been acting on the Deepwater Horizon and on other rigs. The incentives, the steep gradient of profits and career advancement, remain in effect in all fields. The trend of looking for oil in dangerous places with complex technology is, for now, irreversible. Had the accident not happened on the Deepwater Horizon, it might well have happened on some other rig. The Deepwater Horizon was a normal accident, a system accident. Oil companies can and will correct the things that caused this blowout, but complex technologies have other ways of failing that humans cannot foresee. The probability of similar accidents may now be reduced, but it can be reduced to zero only when declining EROEI makes deep-sea production energetically unprofitable.

It is fashionable to think that we will be able to produce renewable energy with gentler technologies, with simpler machines that produce less damage to the earth, the atmosphere, and people. We all hope so, but we must approach such technologies with a dose of realism and a long-term perspective. A geothermal energy project in Basel, Switzerland, begun in December 2006, had been underway only a few days when there was a small earthquake of magnitude 3.4, frightening people and causing minor damage. More than 100 aftershocks continued into 2007, and the project was abandoned because people were too scared. Solar and wind power, at a scale large enough to be meaningful, would consume great amounts of land, and the plants and animals that these lands support. Wind power also kills birds. (Although some people find this troubling, wind power kills but a tiny fraction of the birds killed each year by automobiles and cats.) To use tidal energy on a large scale could mean that large estuaries and stretches of treasured coastlines would need to be engineered to become industrial environments. Renewable energy that gives the same power per person as we enjoy today would not be free of environmental damage. Indeed, in the large land areas that it would require, renewable energy could cause more environmental damage than that caused by our use of fossil fuels. We know that this is not a pleasant observation, but throughout this book we have emphasized the need for realism.

What are the alternatives? The fiscal crises currently experienced by many governments give us a taste of what would lie in store for us should our energy sources ever prove inadequate. Today's fiscal crises are caused by the meltdown of house prices and our financial system (weaknesses that may themselves have been exacerbated by the high oil prices of 2008). A fiscal crisis similar or worse would result if energy shortages were to cause an economic contraction of, say, 5%. What are those consequences? Colorado Springs has been forced to turn off one third of its streetlights. Local governments are unpaving roads that they cannot maintain and returning them to gravel. Teachers are being laid off, programs canceled, and school years shortened. Britain is eliminating whole agencies of government, and planning to implement the most drastic curtailment of public services since World War II. All this is happening at a time when energy is still abundant and relatively inexpensive.

Our societies cannot postpone a public discussion about future energy. As we stated earlier, this must be an adult discussion, a discussion that is honest, serious, and realistic. It cannot be grounded in punditry, or faith-based economics, or unlimited technological optimism. Citizens must demand that politicians and journalists lead these discussions in a spirit of concern for our common future. Few politicians practicing today have shown the capacity to do so. We need a holistic understanding of how energy and society coevolve over the long term, and of the energy–complexity spiral. Thinking that is holistic and long-term is the best way to approach our future, but humans do not normally think this way. We believe, however, that such thinking can be taught if there is the will to do so. Parents and educators should teach children, starting at an early age and on a continuing basis, to respect energy as rare and precious, and to think about how we will use energy long into the future. The earlier we make a serious effort to plan new sources of energy, the easier the transition to future energy will be. And conversely, the longer we continue to delay and obfuscate a discussion of the future, the more wrenching we and our children will find that future when it arrives. We can anticipate and plan for our future, or we can simply allow the future to happen. This is our choice.

The era of plentiful petroleum will someday end, we hope without any more accidents of the magnitude of the Deepwater Horizon blowout. We don't know when this will happen, nor does anyone else. Surely it will happen sooner than we want. Yet we are not without some ability to understand how the future will unfold. We can project the future based on past experience, for we are not the first people to encounter challenges of energy. Always in our discussions it is worthwhile to keep in mind the restatement of Stein's Law: A trend that can't continue, won't.

Further Reading

Complexity

1. Allen, T.F.H., Tainter, J.A., Hoekstra, T.W.: Supply-Side Sustainability. Columbia University Press, New York (2003)
2. Barstow, D., Rohde, D., Saul, S.: Deepwater Horizon's final hours. The New York Times, 25 Dec 2010. http://www.nytimes.com/2010/12/26/us/26spill.html
3. Tainter, J.A.: The Collapse of Complex Societies. Cambridge University Press, Cambridge (1988)

Circumventing the Energy-Complexity Spiral (or not)

4. Jevons, W.S.: The Coal Question: An Inquiry Concerning the Progress of the Nation and the Probably Exhaustion of Our Coal-Mines, 2nd edn. Macmillan, London (1866)
5. Owen, D.: The efficiency dilemma: if our machines use less energy will we just use them more? The New Yorker 86, 78 (2010)
6. Potter, A.: Planet-friendly design? Bah, Humbug. Macleans.Ca. http://www.macleans.ca/article.jsp?content=20070202_154815_4816. Accessed 22 Jan 2011
7. Strumsky, D., Lobo, J., Tainter, J.A.: Complexity and the productivity of innovation. Systems Research and Behavioral Science. 27, 496–509 (2010)

Energy Returned on Energy Invested and Renewable Energy

8. Cleveland, C.J.: Net energy from the extraction of oil and gas in the United States. Energy 30, 769–782 (2005)

9. Hall, C.A.S., Powers, R., Schoenberg, W.: Peak oil, EROI, investments and the economy in an uncertain future. In: David, P. (ed.) Biofuels, Solar and Wind as Renewable Energy Systems: Benefits and Risks. Springer, New York (2008)
10. Hall, C.A.S., Balogh, S., Murphy, D.J.R.: What is the minimum EROI that a sustainable society must have? Energies **2**, 25–47 (2009). http://www.mdpi.com/journal/energies
11. Kubiszewski, I., Cleveland, C.J., Endres, P.K.: Meta-analysis of net energy return for wind power systems. Renewable Energy **35**, 218–225 (2009)
12. MacKay, D.J.C.: Sustainable Energy—Without the Hot Air. UIT Cambridge, Cambridge (2009)
13. Tainter, J.A., Allen, T.F.H., Little, A., Hoekstra, T.W.: Resource transitions and energy gain: contexts of organization. Conservation Ecology **7**(3), 4 (2003), http://www.consecol.org/vol7/iss3/art4

The Future

14. Ackerman, F., Stanton, E.A.: The Cost of Climate Change: What We'll Pay if Global Warming Continues Unchecked. Natural Resources Defense Council, New York (2008)
15. Daly, H.E.: Steady-State Economics: the Economics of Biophysical Equilibrium and Moral Growth. W. H. Freeman, San Francisco (1977)
16. Fridley, David: Nine Challenges of Alternative Energy. Post Carbon Institute, Santa Rosa (2010). http://www.postcarbon.org/Reader/PCReader-Fridley-Alternatives.pdf. Accessed 22 Jan 2011
17. Kemfert, C., Schumacher, K.: Costs of inaction and costs of action in climate protection: assessment of costs of inaction or delayed action of climate protection and climate change. In: Höppe, P., Pielke, R. (eds.) Workshop on Climate Change and Disaster Losses: Understanding and Attributing Trends and Projections, pp. 149–180, Hohenkammer, Germany. http://cstpr.colorado.edu/sparc/research/projects/extreme_events/munich_workshop/full_workshop_report.pdf. Accessed 22 Jan 2011

Proximate and Ultimate Causes

18. Perrow, C.: Normal Accidents: Living with High-Risk Technologies. Basic Books, New York (1984) (Second edition, 1999, published by Princeton University Press)

Appendix A

Glossary

Many of the definitions below are abbreviated from the Oilfield Glossary by Schlumberger.

Annulus is the space between two concentric objects, such as between the wellbore and casing or between casing and tubing, where fluid can flow.

Barrel (bbl) has the volume of 42 U.S. gallons, or 5.615 cubic feet, or 159 liters. Oil and water production produced each day are measured in barrels per day (bpd).

Booster line or mud booster line provides additional mud to the base of the drilling riser to maintain return fluid velocity.

BOP is an acronym for Blowout Preventer stack. A typical stack might consist of one to six ram-type preventers and, optionally, one or two annular-type preventers ("annulars"). A typical stack configuration has the ram preventers on the bottom and the annular preventers at the top. The configuration of the stack preventers is optimized to provide maximum pressure integrity, safety and flexibility in the event of a well control incident. For example, in a multiple ram configuration, one set of rams might be fitted to close on 5-inch diameter drillpipe, another set configured for 4 1/2-inch. drillpipe, a third fitted with blind rams to close on the open hole and a fourth fitted with a shear ram that can cut and hang off the drillpipe as a last resort. It is common to have an annular preventer or two on the top of the stack because annulars can be closed over a wide range of tubular sizes and the openhole, but are typically not rated for pressures as high as ram preventers.

J.A. Tainter and T.W. Patzek, *Drilling Down: The Gulf Oil Debacle and Our Energy Dilemma*, DOI 10.1007/978-1-4419-7677-2,
© Springer Science+Business Media, LLC 2012

Bottomhole assembly comprises the lower portion of the drillstring, consisting of (from the bottom up in a vertical well) the bit, bit sub, stabilizers, drill collar, heavy-weight drillpipe, jarring devices ("jars"), and crossovers for various thread forms. The bottomhole assembly must provide force for the bit to break the rock (weight on bit), survive a hostile mechanical environment and provide the driller with directional control of the well.

Choke is a device incorporating an orifice that is used to control fluid flow rate or downstream system pressure.

Choke line is a high-pressure pipe leading from an outlet on the BOP stack to the backpressure choke and associated manifold. During well-control operations, the fluid under pressure in the wellbore flows out of the well through the choke line to the choke, reducing the fluid pressure to atmospheric pressure. In floating offshore operations, the choke and kill lines exit the subsea BOP stack and then run along the outside of the drilling riser to the surface.

Drilling fluid or drilling mud is a liquid and gaseous fluid or a mixture of fluids and solids used in operations to drill boreholes into the earth. Classification of drilling fluids has been attempted in many ways, often producing more confusion than insight. One classification scheme, given here, is based only on the mud composition by singling out the component that clearly defines the function and performance of the fluid: (1) water-base, (2) nonwater-base, and (3) gaseous (pneumatic).

Drillpipe is a tubular steel conduit fitted with special threaded ends called tool joints. The drillpipe connects the rig surface equipment with the bottomhole assembly and the bit, both to pump drilling fluid to the bit and to be able to raise, lower, and rotate the bottomhole assembly and bit.

Drillstring is a combination of the drillpipe, the bottomhole assembly, and any other tools used to make the drill bit turn at the bottom of the wellbore.

GOM is an acronym for the Gulf of Mexico.

Jar is a mechanical device used downhole to deliver an impact load to another downhole component, especially when that component is stuck. There are two primary types, hydraulic and mechanical jars. While their respective designs are quite different, their operation is similar. Energy is stored in the drillstring and suddenly released by the jar when it fires. The principle is similar to that of a carpenter using a hammer.

Kelly bushing is an adapter that connects the rotary table to the kelly, a specially profiled pipe that turns the drillstring. The kelly bushing has an inside diameter profile that matches that of the kelly, usually square or hexagonal. It is connected to the rotary table by four large steel pins that fit into mating holes in the rotary table. The rotary motion from the rotary table is transmitted to the bushing through the pins, and then to the kelly itself through the square or hexagonal flat surfaces between the kelly and the kelly bushing. The kelly then turns the entire drillstring because it is screwed into the top of the drillstring itself.

Kill line is a high-pressure pipe leading from an outlet on the BOP stack to the high-pressure rig pumps. During normal well control operations, kill fluid is pumped through the drillstring and annular fluid is taken out of the well through the choke line to the choke, which drops the fluid pressure to atmospheric pressure. If the drillpipe is inaccessible, it may be necessary to pump heavy drilling fluid in the top of the well, wait for the fluid to fall under the force of gravity, and then remove fluid from the annulus. In such an operation, although one high pressure line would suffice, it is more convenient to have two. In addition, this provides a measure of redundancy for the operation. In floating offshore operations, the choke and kill lines exit the subsea BOP stack and run along the outside of the riser to the surface. The volumetric and frictional effects of these long choke and kill lines must be taken into account to properly control the well.

LMRP is an acronym for Lower Marine Riser Package. LMRP is a device to flexibly connect and quickly disconnect the drilling riser from the BOP stack.

MMS is an acronym for the Minerals Management Service, currently BOEMRE, the Bureau of Ocean Energy Management and Regulatory Enforcement.

OCS is an acronym for the Outer Continental Shelf.

Oil equivalent is a unit of energy based on the approximate energy released by burning one barrel of crude oil. The U.S. Internal Revenue Service defines it as equal to 5.8×10^6 BTU (6.1 giga Joules or 1.7 MWh). Natural gas and any crude oil volumes measured at standard conditions are converted to an equivalent volume of a standard crude that contains the same energy. This conversion is useful in comparing different sources of oil and gas production. For example, one barrel of oil equivalent has the energy contained in 5,653 standard cubic feet of typical natural gas or 5.653 mscf. A rule of thumb is that one barrel of oil is equal in energy to 6 mscf of natural gas.

PPG or pounds per gallon is a unit of fluid density used in drilling. 1 lbm/gal = 1 ppg = 0.1199 kg/liter. Distilled water at room temperature has the density of 8.3 ppg. Seawater is salty and denser, and its density is about 8.6 ppg at surface conditions.

Riser or drilling riser is a large-diameter pipe that connects the subsea BOP stack to a floating surface rig to take mud returns to the surface. Without the riser, the mud would simply spill out of the top of the stack onto the seafloor. The riser is a temporary extension of the wellbore to the surface.

Spacer is any liquid used to physically separate one special-purpose liquid from another during drilling and completion operations. Special-purpose liquids, such as cement slurries, are typically prone to contamination, so a spacer fluid compatible with each is used between the two. Spacers are used primarily when changing mud types and to separate mud from cement during cementing operations. In the former, an oil-base mud must be kept separate from a water-base mud. In the latter operation, a chemically treated water spacer usually separates drilling mud from cement slurry.

Standard cubic foot (scf) is the volume of gas measured at standard conditions of temperature (59°F) and pressure (1 atmosphere, 101,325 Pascales, or 14.696 pounds per square inch absolute, psia).

Stinger is a specially profiled probe attached to the bottom of a tubing string, designed to engage another mechanical downhole device during a drilling-related operation.

Wiper plug is a rubber-bodied, plastic- or aluminum-cored device used to separate cement and drilling fluid as they are being pumped down the inside of the casing during cementing operations. A wiper plug also removes drilling mud that adheres to the inside of the casing, diminishing the possibility of cement contamination.

Appendix B

Offshore Production

First we describe the different types of offshore installations that are used to produce oil and gas in the Gulf of Mexico. Depending on the water depth and design, offshore platforms are classified[1] as follows.

Compliant towers are slender flexible towers on top of a pile foundation. These platforms support a conventional deck for drilling and production operations. Compliant towers are designed to sustain significant lateral deflections and forces, and are typically used in water depths ranging from 1,500 to 3,000 feet (450–900 meters).

Drillships are maritime vessels fitted with drilling equipment, often used in exploratory drilling of oil or gas wells in deep water and in scientific drilling. Early versions were built on a modified tanker hull, but purpose-built designs are used today. Most drillships are outfitted with a dynamic positioning system to maintain position over the well. They can drill in water depths up to 12,000 feet (3,660 meters).

Fixed platforms are built on concrete and/or steel legs anchored directly onto the seabed, supporting a deck with space for drilling rigs, production facilities and crew quarters. Such platforms are immobile and designed for very long lives. Figure B.1 shows an example of a fixed platform in the Gulf of Mexico. Various types of structure are used, steel jacket, concrete caisson, floating steel, and even floating concrete. Steel jackets are vertical sections made

---

[1] See oilfielddirectory.com/oilfield/oil_platform.htm.

J.A. Tainter and T.W. Patzek, *Drilling Down: The Gulf Oil Debacle and Our Energy Dilemma*, DOI 10.1007/978-1-4419-7677-2,
© Springer Science+Business Media, LLC 2012

Fig. B.1 A fixed platform in the Gulf of Mexico. (Source: www.digitaljournal.com)

of tubular steel members, and are usually piled into the seabed. Concrete caisson structures, pioneered by the Condeep concept, often have in-built oil storage in tanks below the sea surface and these tanks were often used as a flotation capability, allowing them to be built close to shore (Norwegian fjords and Scottish firths are popular because they are sheltered and deep enough) and then floated to their final position where they are sunk to the seabed. Fixed platforms are economically feasible for installation in water depths up to about 1,700 feet (520 meters).

Floating production systems are mainly FPSOs (floating production, storage, and offloading systems), FSO (floating storage and offloading systems), and FSU (floating storage units). FPSOs are large ships equipped with

Fig. B.2 Mærsk Inspirer, one of two identical rigs sharing the title of being the world's largest and most advanced harsh environment jack-up rigs. (Source: www.maersk-drilling.com)

processing facilities. They are moored to a location for extended periods and do not drill for oil or gas.

Jack-ups can be jacked up above the sea using three or four legs that can be lowered like jacks. These platforms are typically used in water depths up to 400 feet (120 meters), although some designs can go to 550 feet (170 meters) depth. They are designed to move from place to place, and then anchor themselves by deploying the legs to the ocean bottom using a rack and pinion gear system on each leg. A jackup rig is shown in Fig. B.2. There are more jackup rigs worldwide than any other type of mobile offshore drilling structures. Specialized offshore drilling contractors deployed the first jackup rigs for use in 199- to 400-feet water depths in the Gulf of Mexico.

Semisubmersible platforms have hulls (columns and pontoons) of sufficient buoyancy to cause the structure to float, but of weight sufficient to keep the structure upright. Semisubmersible platforms can be moved from place to place, and can be ballasted up or down by altering the amount of flooding in their buoyancy tanks. They are generally anchored by combinations of chain, wire rope and/or polyester rope during drilling and/or production operations, although in deep water they can also be kept in place by the use of dynamic positioning. Semisubmersibles can be used in water depths from 200 to 10,000 feet (60–3,050 meters). More generally, a semisubmersible is a specialized marine vessel with good stability and seakeeping characteristics. The semi-submersibles are commonly used as offshore drilling rigs, safety vessels, oil production platforms, and heavy lift cranes. The first "semisub," as they are called, appeared by accident in 1961. Blue Water Drilling Company owned and operated the four-column submersible Blue Water Rig No.1 in the Gulf of Mexico for Shell Oil Company. As the pontoons were not sufficiently buoyant to support the weight of the rig and its equipment, it was towed between locations at a draft[2] midway between the top of the pontoons and the underside of the deck. It was observed that the motions at this draft were very small and Blue Water Drilling and Shell jointly decided that the rig could be operated in the floating mode. The first purpose-built drilling semisubmersible Ocean Driller was launched by ODECO in 1963. Since then, many semisubs have been built, with the Deepwater Horizon rig, shown in Fig. B.3, being a supersized, supercomplicated descendant of this lineage.

Spar platforms are moored to the seabed like the TLPs, but whereas the TLPs have vertical tension tethers, the Spar has more conventional mooring lines. Spars have been designed in three configurations: the "conventional" one-piece cylindrical hull, the "truss spar" where the midsection is composed of truss elements connecting the upper buoyant hull (called a hard tank) with the bottom soft tank containing permanent ballast, and the "cell spar" which is built from multiple vertical cylinders. The Spar may be more economical to build for small- and medium-sized rigs than the TLP, and has more inherent stability than a TLP because it has a large counterweight at the bottom and does not depend on the mooring to hold it upright. It also has the ability, by use of chain-jacks attached to the mooring lines, to move horizontally over the oil field. The first production spar was Kerr McGee's Neptune, which is a floating production facility anchored in 1,930 feet (588 meters) in the Gulf of Mexico. Eni's Devil's Tower is located in 5,610 feet (1,710 meters) of water, in the Gulf of Mexico. Shell's Perdido Spar operates in 8,000 feet (2,438 meters) of water, also in the Gulf.

Tension-leg platforms or TLPs are floating platforms tethered to the seabed in a manner that eliminates most vertical movement of the structure. TLPs

---

[2] The depth of water a ship draws especially when loaded.

Fig. B.3 The Deepwater Horizon was a semi-submersible drilling rig capable of operating in harsh environments and water depths up to 8,000 feet (upgradeable to 10,000 feet) using 18-3/4 inch. 15,000 psi blowout preventer (BOP) and 21 inch. OD marine riser. The rig could drill to the maximum water depth of 30,000 feet/9,144 meters. It displaced 8,000 metric tonnes of water and was 396 feet/121 meters long, 256 feet/78 meters wide, and 136 feet/41 meters deep. (Source: Transocean, http://www.nytimes.com/2010/12/26/us/26spill.html)

are used in water depths up to about 6,000 feet (2,000 meters). The "conventional" TLP is a four-column design which looks similar to a semisubmersible. Mini TLP are relatively low cost units used in water depths between 600 and 4,300 feet (200 and 1,300 meters). Mini TLPs are also be used as utility, satellite or early production platforms for larger deepwater discoveries.

Unmanned installations (sometimes called toadstools), are small platforms, consisting of little more than a well bay, helipad, and emergency shelter. They are designed to operate remotely under normal conditions, only to be visited occasionally for routine maintenance or well work.

Terms such as "jackups," "semisubs," "drillships" describe more than 620 mobile offshore drilling rigs and barges that are available for service anywhere on Earth.[3] The world's deepest platform is currently the Perdido in the Gulf

---

[3] Source: www.rigzone.com/data/

Fig. B.4 The Remotely Operated Vehicle (ROV) Hercules equipped with high-intensity lights at the top, a high-definition video camera in the center, and manipulator arms, including one arm with force feedback, giving an operator the "feel" of handling delicate specimens miles below the ocean's surface. Similar ROV's did most of the work necessary to secure the failed Macondo well. In routine operations, such ROV's are used to construct and repair seafloor equipment and installations. Without them ultra-deepwater field operations would be impossible. (Source: NOAA, the photo is courtesy of Mystic Aquarium/IFE)

of Mexico, floating in 8,000 feet/2,438 meters of water, and designed for 100,000 barrels of oil equivalent per day. It is operated by Royal Dutch Shell and was built at a cost of $3 billion.

Once wells are drilled and completed, they need to be attached to a platform or a ship. The hydrocarbons these wells produce must then be separated from water and delivered to shore, where they can be processed into petroleum and gas products.

Almost all subsea construction and repair work is done by the remotely operated vehicles (ROVs), similar to the one shown in Fig. B.4. For a couple of months, you could watch real-time video feeds from the subsea operations by these ROVs in the aftermath of the Macondo well blowout. Without the robotic technology all work at water depths below 100 meters would be

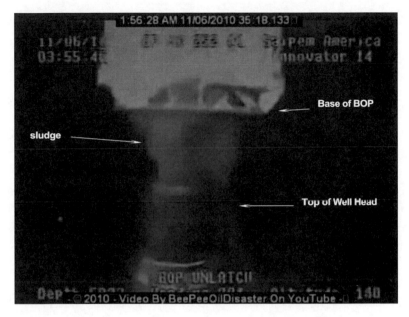

Fig. B.5 Live video feed from an ROV shows the unlatched Blowout Preventer stack being lifted from the wellhead of the killed Macondo well. (Source: A frame from the BP live video feed, 11/06/2011)

impossible. Watching live video feeds from the various activities at the Macondo well became somewhat of an obsession to many of us. Figure B.5 shows what we hope is one of the final clips from the Macondo well reality show.

Now, we give you a few glimpses of the incredible engineering and cutting-edge technology that make tens of thousands of offshore wells work without a hitch, day-after-day for years, and deliver the much-needed oil and gas to consumers. All but one of the offshore production systems sketched below are described in far more detail in a report by five engineering students from Brazil that is emerging as a superpower of offshore engineering.[4] As it turns out, engineering education is quite relevant to the Macondo well tragedy.

---

[4] Tiago Pace Estefen, Daniel Santos Werneck, Diogo do Amaral Macedo Amante, João Paulo Carrijo Jorge, and Leandro Cerqueira Trovoado, *A Subsea Production System for Gas Field Offshore Brazil*, Federal University of Rio de Janeiro Naval Architecture and Ocean Engineering Department, International Student Offshore Design Competition, 2005. Petrobras, the publicly-traded Brazilian oil company, is on its way to becoming larger than Exxon-Mobil.

FPSOs

The new MMS just gave Petrobras a permit to operate a new floating production, storage and offloading (FPSO) in the deep Gulf of Mexico. An FPSO ship shown in Fig. B.6 is used to process and store oil and gas on the high sea far from shore and away from pipelines. An FPSO vessel is designed to receive oil and/or gas produced from nearby drilling or production platforms, or subsea terminals, and process and store them until offloading onto a tanker. In 2009, Shell and Samsung announced an agreement to build up to ten Liquified Natural Gas (LNG) FPSOs. Their likely size and capacity are 456 meters in length by 74 meters in width and 450,000 cubic meters, respectively. They will cost an estimated $5 billion. Talk about complexity!

An FPSO can be a converted oil tanker, but it usually is custom designed and built. A vessel used to store oil only is referred to as a floating storage and offloading vessel (FSO). The oil/gas production lines are connected to an area of the ship called a "turret." The turret can be external and hanging off the side of a barge- or platform-like FPSO in calmer waters of West Africa. In harsh environments like the North Sea, an internal turret is located in the center and underneath the FPSO and the vessel is built like a ship.

Floating production, storage, and offloading vessels are particularly effective in remote or deepwater locations where seabed pipelines are not cost effective.

Fig. B.6 A floating production, storage and offloading (FPSO) ship on the left is connected to a tanker on the right. The FPSO ship processes and stores production from an offshore installation and off-loads periodically to tanker ships. (The image is courtesy of Bluewater Energy Services B.V., The Netherlands, www.bluewater.com)

Fig. B.7 Eight subsea wells are connected to a semisubmersible platform, which in turn is connected to a pipeline exporting hydrocarbons to shore. (Source: NOAA, the photo is courtesy of Mystic Aquarium/IFE)

FPSOs eliminate the need to lay expensive long-distance pipelines from the oil (and perhaps gas) wells to an onshore terminal. They can also be used economically in smaller fields that are exhausted in a few years and do not justify expensive production platforms. Once the hydrocarbons are depleted, the FPSO can be moved to a new location. In areas of the world subject to rough seas and hurricanes (the North Sea and the Gulf of Mexico), cyclones (northwestern Australia), or icebergs (Canada), some FPSOs are able to release their mooring/riser turret and steam away to safety in an emergency.

Semisubmersibles

The subsea production system shown in Fig. B.7 consists of eight wells on the seafloor connected to a floating platform by flowlines and thermally insulated flexible risers. Thermal insulation and/or electrical heating and/or injection of methanol or glycol into flowlines are necessary to prevent the formation of methane hydrates that would plug these flowlines just as they plugged the unfortunate cofferdam used by BP to capture oil and gas erupting

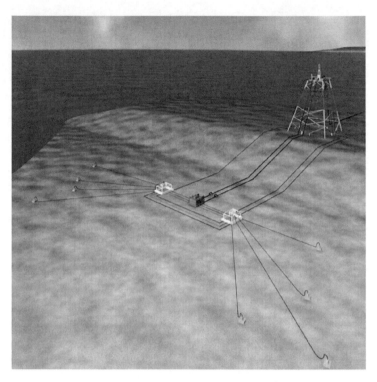

Fig. B.8 A fixed platform-subsea production system. The two production manifolds are shown in white. The Pipeline End Manifold (PLEM) is in the middle. (Source: NOAA, the photo is courtesy of Mystic Aquarium/IFE)

from the Macondo well. The oil/gas mixture is dehydrated in the process plant on the platform, compressed, and exported[5] through a hybrid riser to the right of the platform. Water from the separation process is treated and disposed of into the sea. The hybrid riser consists of three flexible risers of small diameter linking the platform to a bulky vertical rigid riser. The vertical riser is connected to a pipeline that exports hydrocarbons directly to shore.

Fixed Platforms

The fixed platform subsea production system in Fig. B.8 consists two parallel manifolds, each connected to four wells through flexible flowlines. Each manifold has two production headers connected to a Pipeline End Manifold (PLEM) in order to allow hydrocarbons to flow through two rigid pipes up to the platform.

---

[5] The gas is often flared.

The left manifold receives an umbilical and the right one receives a service flexible pipe and another umbilical. Umbilicals are large-diameter flexible cables that remotely control the subsea equipment. Umbilicals transfer hydraulic pressure and electrical power to operate submerged equipment and valves. They also send control commands and retrieve sensor data through electrical and/or optical fiber cables. Umbilicals can contain additional hoses for the injection of chemicals, such a mono-ethylene glycol or methanol.

An electro-hydraulic control system is used because of the distance between the platform and the wells (10–20 kilometers), enabling faster valve activation in comparison with a hydraulic only system (a similar electro-hydraulic system operated the Macondo well Blowout Preventer stack). In the process plant on the platform, water associated with the hydrocarbons is removed before compression and exported through a rigid riser to shore.

Subsea-to-Shore

The subsea-to-shore configuration in Fig. B.9 is identical to the one described above, but without using a fixed platform. Hydrocarbon transport to an onshore terminal is accomplished with two large-diameter, rigid pipes. The production pipes are thermally insulated. During a well shutdown, glycol must be injected into the flow lines, and one of the umbilicals has an internal hose for chemical injection. The wells are far from the shore, about 100 miles or 160 kilometers. Therefore, again, a multiplex electro-hydraulic control system is used. As the reservoir pressure declines with time, it may be necessary to install subsea separators and gas compressors to guarantee production at a cost of an even more complex submerged system.

Fig. B.9 A sea-to-shore production system does not need a fixed platform. (Source: NOAA, the photo is courtesy of Mystic Aquarium/IFE)

Appendix C

Operating an Offshore Platform

A brief list of essential personnel on an offshore rig includes the following:

Ballast control operator controls fire and gas systems.

Catering crew are tasked with performing essential functions such as cooking, laundry, and cleaning the accommodation.

Control room operators are in charge of controlling FPSO vessels or production platforms.

Coxwains maintain the lifeboats and man them if necessary.

Crane operators are in charge of the cranes for lifting cargo around the platform and between boats.

Dynamic positioning operator is responsible for navigation, ship or vessel maneuvering for mobile offshore drilling units (MODU), station keeping, and fire and gas systems operations in the event of incident.

Helicopter pilot(s) lives on some platforms and transports workers to other platforms or to shore upon crew changes.

Maintenance technicians are responsible for the proper workings of instruments and sensors, and electrical and mechanical systems.

On the Deepwater Horizon semisubmersible rig there was apparent lack of clarity regarding who was in charge, the OIM or the Captain. Such a lack of clarity can be very damaging during emergencies and accidents.

J.A. Tainter and T.W. Patzek, *Drilling Down: The Gulf Oil Debacle and Our Energy Dilemma*, DOI 10.1007/978-1-4419-7677-2, © Springer Science+Business Media, LLC 2012

Offshore installation manager (OIM) is the ultimate authority during his shift and makes the essential decisions regarding the operation of the platform.

Offshore operations engineer (OOE) is the senior technical authority on the platform.

Operations coordinator manages crew changes.

Operations team leader (OTL) is responsible for coordinating operational inputs to a project, ensuring that the operations are safe, reliable, effective, and efficient.

Production technicians operate, for example, potable water system, service water system, reverse osmosis unit, jet kill pump, air system, glycol system, diesel system, quarters systems, equipment and HVAC, fuel gas system, turbine engines, reciprocating engines, power generation and distribution, corrosion control, valves, wireline, documents and drawings, radio communication systems, safe purging and pressure testing, and so on.

Scaffolders rig up scaffolding for when it is required for workers to work at height.

Second mate meets manning requirements of flag state, operates fast rescue craft, cargo operations, and is a fire team leader.

Third mate meets manning requirements of flag state, operates fast rescue craft, cargo operations, and is a fire team leader.

During drilling, logging, and/or well workover operations, there are extra people on the rig or platform, for example:

Company Man represents an operating/exploration company. He is also known as company representative, foreman, drilling engineer, company consultant, rig site leader, or well site manager. It appears that during the drilling of the Macondo well, the BP Company Man often overruled what now seem to be not-so-strenuous objections of the Halliburton and/or Transocean employees.

Derrickman or derrickhand is one position below the driller and almost always reports directly to the driller. The name derrickman comes from the location at the top of the derrick he normally occupies. From this location he guides the stands of drill pipe, typically 90 feet (27 meters) long, into the fingers at the top of the derrick while tripping (removing the drill string) out of the hole. When tripping into the hole he pulls the pipe out of the fingers and guides it into the top drive or the elevators. Traditionally, the derrickman works closely with the mud engineer when he is not tripping pipe and he is not needed at the derrick. In this capacity, it is derrickman's responsibility to monitor the mud viscosity and weight (density), add sacks of chemicals (25–100 pounds each) to the mud to maintain properties, and monitor the mud level in the mud pits to aid in the well control. Sometimes, he may be responsible for the shale shakers and the mud pumps. The derrickman is also responsible for the transfer of chemicals (barite or bentonite, or oil-based fluids) from bulk silos or tanks to the mud system. The Deepwater Horizon's derrickman did not or could not observe the changes in mud level during the critical negative pressure test.

Driller is a team leader (superintendent) in charge of well-drilling operations. The driller is responsible for interpreting the symptoms of high gas and fluid pressure. In an emergency he is also responsible for taking the correct counter-measures to stop an uncontrolled well response (a pressure kick or blow out) from emerging. The driller watches for gas levels coming out of the hole, how much drilling mud is going in and out, and for any other information pertinent to well drilling. In offshore operations the driller is in charge of real-time decisions. In the failed negative pressure test of the Macondo well the driller was responsible for real-time responses to the sensor information and visual observations of the well flowing back.

Mud Engineer or **Drilling Fluids Engineer**, but most often referred to as the "Mudman," is responsible for ensuring that the properties of the drilling fluid, also known as drilling mud, are within required and often changing specifications.

Logging witness is the leader of the logging service project team consisting of the service company logging engineer, his/her crew and the drilling superintendents on the rig.

Roughnecks work on the drill floor of a drilling rig handling specialized drilling equipment for drilling and pressure controls. In practice, roughnecks range from unskilled to highly skilled, depending on the individual worker's aptitude and experience.

Roustabouts perform general labor, such as loading and unloading cargo from crane baskets and assisting welders, mechanics, electricians, and other skilled workers.

Toolpusher is in charge of the drilling department and reports to Captain or Offshore Installation Manager (OIM).

About the Authors

Joseph A. Tainter is Professor of Sustainability in the Department of Environment and Society, Utah State University, having previously served as Department Head. He received his Ph.D. in Anthropology from Northwestern University in 1975. Dr. Tainter worked on issues of sustainability before the term became common, including his highly-acclaimed book *The Collapse of Complex Societies* (Cambridge University Press, 1988). He is co-editor of *The Way the Wind Blows: Climate, History, and Human Action* (Columbia University Press, 2000), a work exploring past human responses to climate change. With T. F. H. Allen and Thomas Hoekstra he wrote *Supply-Side Sustainability* (Columbia University Press, 2003), the first comprehensive approach to sustainability to integrate ecological and social science. Dr. Tainter has taught at the University of New Mexico and Arizona State University. Until 2005 he directed the Cultural Heritage Research Project in Rocky Mountain Research Station. Dr. Tainter's sustainability research has been used in more than 40 countries, and in many scientific and applied fields. Among other institutions, his work has been consulted in the United Nations Environment Programme, UNESCO, the World Bank, the Rand Corporation, the International Institute for Applied Systems Analysis, the Beijer Institute of Ecological Economics, the Earth Policy Institute, and the Technology Transfer Institute/Vanguard. Dr. Tainter has been invited to present his research at the Getty Research Center, the University of Paris (Panthéon-Sorbonne), the Royal Swedish Academy of Sciences, and many other venues. His research has been applied in numerous fields, including economic development, energy, environmental conservation, health care, information technology, urban studies, and the challenges of security in response to terrorism. He appears in the film The 11th Hour, produced by Leonardo DiCaprio, Leila Conners Petersen, Brian Gerber, and Chuck Castleberry, and in the ABC News special Earth 2100. Dr. Tainter's current research focuses on sustainability, energy, and innovation.

Tad Patzek is the Lois K. and Richard D. Folger Leadership Professor and Chairman of the Petroleum and Geosystems Engineering Department at The University of Texas at Austin. He also holds the Cockrell Regents Chair #11. Between 1990 and 2008, he was a Professor of Geoengineering at the University of California, Berkeley. Prior to joining Berkeley, he was a researcher at Shell Development, a unique research company managed for 20 years by M. King Hubbert of the Hubbert peaks. Patzek's current research involves mathematical and numerical modeling of earth systems with emphasis on fluid flow in the subsurface soils and rocks. He works on the thermodynamics and ecology of human survival, especially on the global food and energy supply systems. More recently, Patzek has engaged in the studies of complex systems, focusing on the ultra deepwater offshore operations. He briefed Congress on the BP Deepwater Horizon well disaster in the Gulf, and was a frequent guest on NPR, ABC, BBC, CNN, and CBS programs. For the last two years, Patzek's research has emphasized the use of unconventional natural gas as a fuel bridge to the possible new energy supply schemes for the U.S. He appeared in the Haynesville Shale documentary. Currently, Patzek teaches courses in petroleum engineering, hydrology, ecology and energy supply, computer science, and mathematical modeling of earth systems. Patzek is a coauthor of over 200 papers and reports, and is writing five books. In March 2011, he was chosen by US Interior Secretary Ken Salazar to serve on the Ocean Energy Safety Advisory Committee, a permanent advisory body providing critical guidance on improving offshore drilling safety, well containment, and spill response offshore.

Index